Global Energy Economics and Climate Protection
Report 2009

W0079386

Valentin Crastan

# Global Energy Economics and Climate Protection Report 2009

 Springer

Dr. Valentin Crastan
ch. des Blanchards 18
2533 Evilard
Switzerland
valentin.crastan@bluewin.ch

ISBN 978-3-642-11872-2     e-ISBN 978-3-642-11873-9
DOI 10.1007/978-3-642-11873-9
Springer Heidelberg Dordrecht London New York

Library of Congress Control Number: 2010923829

*Cover design:* WMXDesign GmbH, Heidelberg

Printed on acid-free paper

Springer is part of Springer Science+Business Media (www.springer.com)

# Preface

At the end of 2009 the World Bank and the IMF published new figures for gross domestic product adjusted for purchasing power. In addition to the new assessment of various developing and emerging nations, the considerable downgrading of the mean purchasing power of China and India is of particular significance.

The present English edition of the book takes this into account. Consequently it is not simply a translation of the German edition, but is also an update, in both the introduction and the main report.

Biel, January 2010                                      Valentin Crastan

# Preface to the German edition

Climate change is becoming more and more evident to everybody and there is an urgent need for action. The resonance resulting from the forthcoming conference in Copenhagen can be taken as a sign that even international politics is taking the subject seriously. The need for a worldwide reduction of $CO_2$ emissions is only sporadically doubted. The required extent of this reduction is also largely clear. However, there is continuing argument over how much contribution to this should be made by the individual countries and regions of the world.

The following report is an attempt to provide a basis for rational discussion, with an analysis of the worldwide energy economy and the associated emissions. Measurable indicators based on the two main aspects, namely energy efficiency and $CO_2$ intensity, should enable the efforts to be judged equitably.

These efforts require local comprehension and the corresponding commitment. This commitment needs to be appropriate for the economic power of the country concerned. However, it must be supported by coordinated international action, primarily by the economically strong countries.


Biel, November 2009                                          Valentin Crastan

# Contents

# Introductory remarks

## 1 Climate protection as necessity

Scientific evidence of climate change is provided by the publications resulting from numerous studies and investigations by climate researchers. The urgency of the need to act implies that the worldwide efforts to improve understanding of the phenomena with ever better physical models and to predict future changes with increasing certainty must be supported.

Despite this, there is constant resistance to the so-called "climate hysteria" and against allegedly "dogmatic mainstream research" [7]. The arguments employed are largely of an emotional type or are based on a superficial understanding of past and present climate changes, for example the pseudo-scientific assertion: "The warming currently taking place can be understood as a recovery from the last minor ice age", which has been refuted several times by scientific findings.

It is gratifying that international politics has begun to take seriously the results of scientific efforts. Various summit meetings, such as the forthcoming one in Copenhagen, are expected to come to agreements on targets that can be accepted both by the industrialised nations and by the emerging and developing countries. This is where the main problem of climate protection is to be found and some basic information and material for thought on the subject are presented here.

## 2 Climate change and climate protection, indicators

The first thing to state is that climate change is primarily caused by $CO_2$ emissions arising in the conversion and consumption of energy and the main effort must therefore be directed here. Of course, other important aspects such as the destruction of rain forests or the emission of methane due to agriculture must also be included.

Today it is recognised almost everywhere in science and politics that the world total emissions of $CO_2$ must be reduced as fast as possible. There are difficulties for the emerging and developing countries resulting from the agreements (such as that of Kyoto) that intend to implement reductions in per-capita $CO_2$ emissions relative to the present (or a past) point in time. For obvious reasons, such reductions are unacceptable for these countries. Their accusation, that the present situation results from the behaviour of the industrialised nations, and that the emerging and developing countries must have the same opportunities to develop, has to be accepted as justified.

Consequently the per-capita $CO_2$ emission value is not a suitable general target for short- and medium-term efforts because it does not take account of countries' differing development states. It makes sense as a long-term goal and 1 t $CO_2$ per capita and year is certainly worth having as a long-term target.

A better grip on the problem results if the benchmark or guideline value is not the $CO_2$ emissions per capita but a value relative to an indicator that is a good representation of the development state of the country. The only quantity that is collected world-wide and meets this requirement to some extent is the gross domestic product (GDP), at purchasing-power parity (World Bank, IMF [9], despite the shortcomings associated with this quantity as an indicator of prosperity. Finding and establishing a better one would be a task for the Guild of Economists.

In 2007, the $CO_2$ emissions value expressed in g $CO_2$/$ purchasing-power-parity GDP was 435 g $CO_2$/year-2007 $. This indicator can be determined as the product of two factors: the energy intensity of the GDP and the $CO_2$ intensity of the energy. The first one characterises the efficiency of the use of energy, and the second one the $CO_2$ sustainability of the energy employed. These factors are of equal significance and fundamental for climate protection.

The following chart shows the world situation in the year 2007 (the most recent available data for the whole world). The relative values of the indicator are particularly significant. Central and south America are the best here, thanks to a significant reliance on water power for electricity generation and an acceptable energy efficiency. On the other hand, China, the Middle East and the countries of the former Soviet Union have very poor values because of the enormous waste of energy in these countries (China has also poor $CO_2$ intensity).

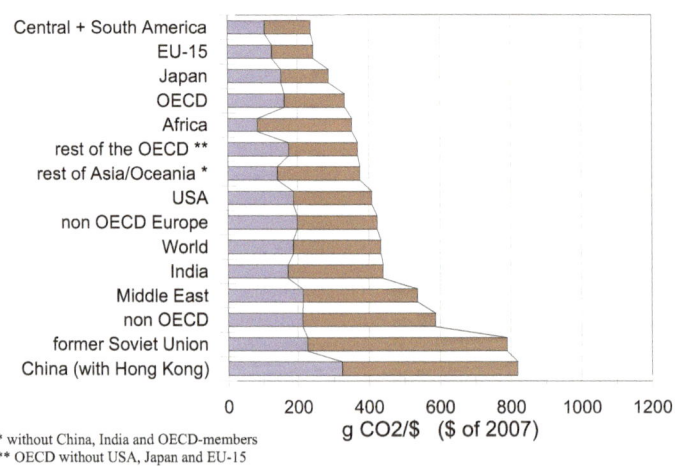

Worldwide $CO_2$ emissions indicator, showing relative significance
of the factors $CO_2$ intensity (blue) and energy intensity (red)

## 3  Target values and international cooperation

For effective climate protection (temperature rise not over 2°C, stabilisation of $CO_2$ emissions by 2030 relative to 2004 and halving by 2050) the following world-wide target values are necessary ([1], main report and annex):

- 3.3 t CO2/capita by 2030, giving about 200 g CO2/$,
- 1.5 t CO2/capita by 2050, giving about 60 g CO2/$.

The worldwide target values in g $CO_2$ per purchasing-power equivalent GDP (year-2007 $) depend on the expected increase in the worldwide GDP and are therefore more difficult to determine than those for $CO_2$ emissions per capita. The development of population can be estimated more easily. Nevertheless they are more informative for comparisons since they take into account the development state of the country concerned. Each country should therefore accept them as a guideline and make an effort to achieve them, regardless of the state of development.

As shown in the above chart, all countries need to catch up, although to varying degrees, if they wish to achieve the goal of about 200 g $CO_2$/$ by 2030. Those countries with a large demographic and political weight have a particularly great responsibility: China, India, USA, European Union, Japan, Brazil, Russia. The means of reaching the target must, in the end, be decided by each country. It will be judged by the results. In view of the significance of electricity generation in the production of $CO_2$ emissions (40-45% worldwide) it must be assumed that a fundamental question is to turn away from today's coal-based economy.

The following chart shows, for the whole world, the annual changes in the indicators (full definitions in the main report) from 2004 to 2007 and the annual percent changes required until 2030 for protecting the climate ($CO_2$ indicator in g $CO_2$/ $ of 2007; details for some regions of the world in the main report, Section 4).

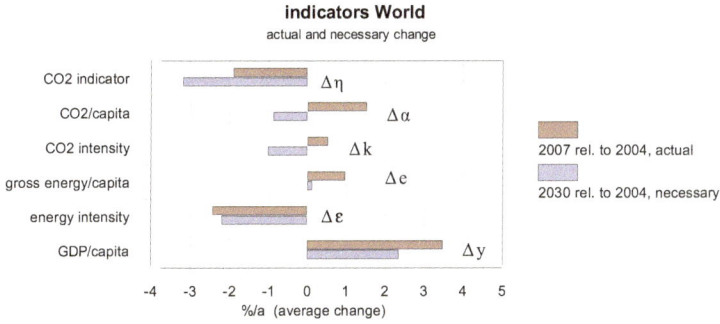

Indicators world: effective annual % change from 2004 to 2007 (red) and worldwide annual % change required until 2030 for climate protection for an assumed growth of GDP (PPP)(blue).
$\varepsilon$ = energy intensity,  k = $CO_2$ intensity, $CO_2$ indicator $\eta = k \cdot \varepsilon$, $\Delta\eta\% = \Delta k\% + \Delta\varepsilon\%$
y = GDP/capita (PPP),  $CO_2$/capita $\alpha = \eta \cdot y$, $\Delta\alpha\% = \Delta\eta\% + \Delta y\%$
gross energy/capita  e $= \varepsilon \cdot y$, $\Delta e\% = \Delta\varepsilon\% + \Delta y\%$

Thus, the $CO_2$ indicator has, according to the above diagram, reduced worldwide by about 2% annually from 2004 to 2007. However, for effective protection of the climate, this indicator must come down by 3.2% annually between 2004 and 2030. If the average growth of GDP(PPP) per capita exceeds the assumed value of just under 2.4%/a, an even greater decrease is required.

With the assumed growth rate of GDP(PPP) the indicators would evolve worldwide from 1971 to 2030 as shown in the following graph.

**Worldwide indicators from 1971 to 2030**
$ of 2007

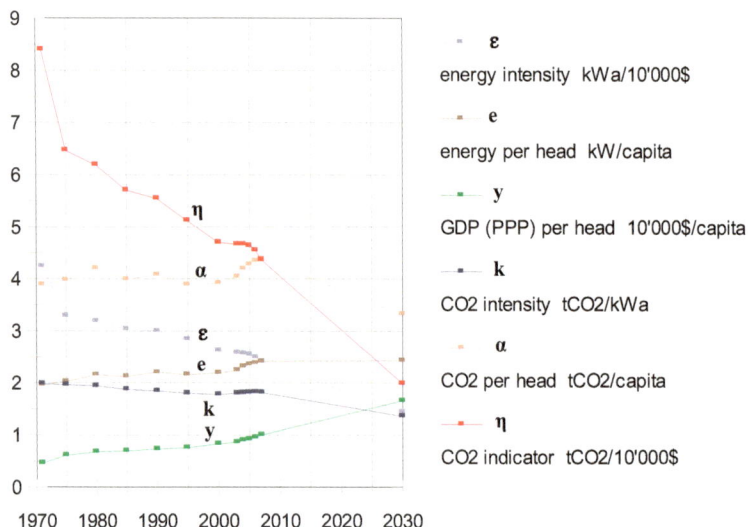

Worldwide indicators from 1971 to 2007 and necessary progress until 2030

International cooperation is certainly required, and naturally both market-based means ($CO_2$ certificates) and promotional actions by the state can and must be employed.

In the following main report the main indicators for all countries are presented and commented. This is together with the hope that critical debate with one's own energy economy and that of the world will give rise to discussion and lead to the understanding that, despite differing and, to some extent conflicting, interests, will enable a coming together of viewpoints in matters of climate protection.

The appendix presents energy economics principles and an analysis of the world's energy situation and the related $CO_2$ emissions that vary with country and sector of the economy (electricity, motor fuels, heating requirements etc.).

# Main report:

# The present world energy situation and implications for climate protection

The crisis in the world of finance and economics is significantly reducing the scope for an ambitious environment policy aimed at containing global warming, but it also opens up some opportunities. In many countries the gross domestic product will decrease, at least for a time. This will result in a temporarily reduced demand for energy, despite price reductions that unfortunately cause false incentives. Thereby, some time is gained in which to consider sensible energy-policy actions and to put them into practice systematically. There is no alternative to climate protection. Particularly in this crisis, it is important for the world to become more aware that the costs of prevention by means of active climate protection will be very much lower than the world-wide damage that could result from global warming. This damage can only partially be avoided by employing solely, and often selfishly, adaptation strategies. It is important to have clearly stated goals and to promote cooperation, especially among the industrialised and emerging nations, but also including the developing countries. Indicators that are reasonable and, above all, accepted, are a prerequisite for negotiations. It is to be hoped that the United Nations Climate Change conference, in Copenhagen in 2009, will create a better understanding of these questions in the industrialised countries and the countries of the third world.

## 1    Objectives and indicators

The climate goals can be formulated relatively easily. Climate studies show that by 2030, the world-wide emission of $CO_2$ by the energy economy must be stabilised at 2004 levels and halved by 2050. This would enable a limitation of the $CO_2$ concentration in the atmosphere to at most 400 ppm and the mean temperature rise of the planet to about 2°C [5], [6].

What limits does this place on the energy economy? The following discussion will start from the situation in the year 2004 [1], [4]. From a world-wide gross energy consumption of 14.9 TWa (1 kWa = 8760 kWh = 0.753 toe, which is the energy of about 1000 litres of petrol) or the rate of 2.35 kW/head, 26 600 Mt $CO_2$ were emitted. This figure has increased in each year since then. In 2007, it was already 29 000 Mt according to IEA [3].

This increase has two causes:

a) Energy consumption increases as a consequence of the world-wide increase in gross domestic product (GDP). This grows with the increase in world population and above all, because of the legitimate increase in prosperity in the emerging and developing countries. The ratio energy/GDP is the *energy intensity* $\varepsilon$ , that can be quantified, for example in kWa/$10 000. To achieve the goals of climate protection requires conversion and use of energy to be as efficient as possible, in order to achieve the lowest possible energy intensity. Hence, the concept of energy efficiency is the reciprocal of energy intensity. In 2004, the world-wide energy intensity, using the values for GDP at purchasing power parity (PPP), was $\varepsilon = 2.57$ kWa/$10 000 (US Dollars of 2007).

b) $CO_2$ emissions from the use of energy are rising because, particularly (but not only) in the emerging nations, there is increasing use of energy sources (oil, coal) that emit large amounts of $CO_2$. This is expressed in the $CO_2$ *intensity k* of the consumed energy in units of t $CO_2$/kWa. It is dependent on the type of energy used. The $CO_2$ released in burning coal gives a value of around $k = 3$ t $CO_2$/kWa, burning oil gives about 2.3 t $CO_2$/kWa and burning natural gas is about 1.7 t $CO_2$/kWa. The use of $CO_2$-free energy sources (water power, solar radiation, wind energy, solar heating, geothermal energy, biomass energy, sea currents and wave power but also nuclear energy) gives a theoretical value $k = 0$. The world-wide average of $CO_2$ *intensity* in 2004 was: $k = 1.78$ t $CO_2$/kWa [1].

This trend of increasing $CO_2$ emissions can only be broken by influencing both of its causes. The important number is the product of the two factors *energy intensity* $\varepsilon$ and $CO_2$ *intensity k, used as $CO_2$ indicator* $\eta$ that must be kept as low as possible for effective climate protection.

$$\eta \left[ \frac{t\,CO_2}{10'000\,\$} \right] = k \left[ \frac{t\,CO_2}{kWa} \right] \cdot \varepsilon \left[ \frac{kWa}{10'000\,\$} \right],$$

The world-wide value of this $CO_2$ indicator in 2004 was about 4.7 t $CO_2$/\$10 000 (US Dollars of 2007) or, written more concisely and understandably, 470 g $CO_2$/\$. This number, together with its components, is significant for comparisons between countries regarding sustainability of their energy economics and the effectiveness of measures for climate protection. It is therefore suitable as a basis for discussion and a starting point for negotiations. What value must it reach in 2030 or 2050 if the goals of climate protection are to be achieved?

The resulting per-capita indicators are also of interest: e for energy and $\alpha$ for $CO_2$ emissions

$$e \left[ \frac{kW}{capita} \right] = \varepsilon \left[ \frac{kWa}{10'000\,\$} \right] \cdot y \left[ \frac{10'000\,\$}{a,\ capita} \right]$$

$$\alpha \left[ \frac{t\,CO_2}{a,capita} \right] = \eta \left[ \frac{t\,CO_2}{10'000\,\$} \right] \cdot y \left[ \frac{10'000\,\$}{a,capita} \right] = k \left[ \frac{t\,CO_2}{kWa} \right] \cdot \varepsilon \left[ \frac{kWa}{10'000\,\$} \right] \cdot y \left[ \frac{10'000\,\$}{a,capita} \right]$$

in which *y* indicates the gross domestic product at purchasing power parity GDP (PPP) per capita. The indicator $\alpha$ is decisive for climate warming since the population trend is the most predictable and it is entirely sensible long-term as a target quantity, but is not particularly suited to current-day comparisons and therefore as a basis for short-term negotiations.

The following will first consider more closely the sustainability of relevant contributing factors $k$ and $\varepsilon$ for the current energy supply situation.

## 1.1   Energy intensity

To compare the energy intensity of the economies of various countries, a measurement unit must be established that enables the economic capacity of a country to be defined in an objective and fair way. The absolute value of the GDP in \$ at the currency's rate of exchange does not seem to be particularly suitable for that because the level of prices and therefore the buying power, that is the real indicator of capacity and competitiveness, can be very different. Therefore international organisations (World Bank, IMF) determine the GDP of all countries at purchasing power parity, the GDP (PPP), [9]. There have also been attempts to define other quantities. However, despite its shortcomings, the GDP (PPP) is the only statistically available characteristic that makes reasonable world-wide comparisons possible.

Extrapolation of this quantity to the year 2030 or 2050 requires scenarios that represent the development of energy consumption corresponding to the growth of population and the GDP. Two scenarios of this type have been developed for 2030 by the IEA (WEO 2004 [2]). They assume that the population of the world will increase from 6.35 bn in 2004 to 8.1 bn in 2030. The GDP (PPP) increases from $57 500 bn in 2004 (expressed in dollars of 2007) to $135 000 bn in 2030, corresponding to a mean rate of increase of 3.3%. Both scenarios, the reference scenario and the IEA alternative scenario, make these assumptions.

The *reference scenario* leads to an energy requirement of 21.9 TWa in 2030, corresponding to a mean rate of increase of 1.5%. The energy intensity would therefore reduce from the 2.57 kWa/$10 000 mentioned above to 61% of that, or 1.58 kWa/$10 000, an efficiency improvement of 63% in the use of energy.

The *alternative scenario*, that takes greater account of the needs of climate protection, leads, thanks to improved efficiency, to an energy requirement of 19.5 TWa, with a mean rate of increase of only 1.1% p.a. The energy intensity would reduce to 1.4 kWa/$10 000, and the efficiency of energy use would increase, compared to 2004, by 83%. World-wide energy consumption per head would rise slightly from 2.35 kW in 2004 to 2.41 kW. The reduction in the industrialised countries (OECD: 6.1 kW/head in 2004) would be more than offset by the increase in the non-OECD world (1.4 kW/head in 2004).

The new *Blue scenario* of the IEA expects an energy demand of 22 TWa for 2050 with an energy intensity of 1 kWa/$10 000.

## 1.2    $CO_2$ intensity

We shall consider only the *alternative scenario*, since the reference scenario is, as regards the goals for climate protection, simply unacceptable. The alternative scenario is not satisfactory either, since it only provides an improvement in $CO_2$ intensity from 1.78 t $CO_2$/kWa in 2004 to 1.73 t $CO_2$/kWa in 2030. Under these conditions the $CO_2$ emissions would not stabilise but, with the above assumptions about gross energy consumption, increase by nearly 30% to 34 000 Mt. The structural change in the energy economy included in the alternative scenario (turning away from the present coal-based economy) is too small to be able to put a real brake on climate change. With this scenario, the above-defined *$CO_2$ indicator* $\eta$ of sustainability would improve insufficiently from 510 g $CO_2$/$ only to 270 g $CO_2$/$. The worldwide $CO_2$ emissions per head would stay at practically the same value as the 2004 figure of 4.2 t $CO_2$ until 2030.

A significant improvement in the $CO_2$ intensity is not expected until after 2030. The *Blue scenario* of the IEA for 2050 sets a target value of 0.8 t $CO_2$/kWa. With the worldwide GDP (PPP) of $216 000 bn (of 2000) expected at that time the worldwide emission of $CO_2$ could be reduced to around 18 Gt and the mean $CO_2$ emission per capita to around 2 t $CO_2$.

## 2  Climate protection and required measures

To satisfy the requirements of climate protection, the assumptions of the IEA alternative scenario, that seem realistic and reasonable as regards efficiency improvement, must be supplemented by a greater structural change in the energy economy in order to achieve a considerably greater improvement in the $CO_2$ intensity figure. To make this possible, it is the duty of national and international politics to set the required general framework that will promote the required investments in the economy, including trade in emissions and perhaps also $CO_2$ taxes. Worldwide knowledge of the indicators is useful for this, as they provide an objective basis for negotiation of appropriate stimuli, directives and agreements.

Before reminding ourselves of the measures required for a reduction of $CO_2$ intensity, the following diagram shows a breakdown of $CO_2$ emissions in 2004 (end users of energy (heat, fuels, electricity) and losses in the energy sector, def. see [1] or annex):

### global CO2 emissions, 2004
in % of total of 26600 Mt

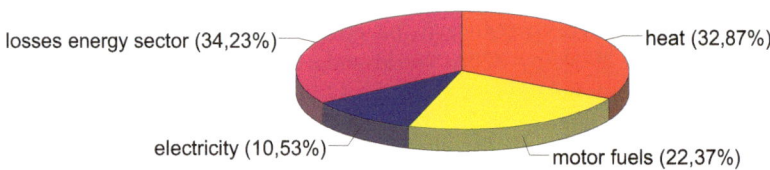

losses energy sector (34,23%)    heat (32,87%)
electricity (10,53%)    motor fuels (22,37%)

It can be seen that about 45% of emissions come from the energy sector that serves mainly for the production of electricity. The worldwide level of electrification corresponds largely to the level of economic development. In addition to actions in the area of heating and mobility, which also partially have the effect of increasing the need for electricity (heat pumps, electric cars), it is therefore of primary importance to generate electricity with the least possible production of $CO_2$. The following are possible ways to achieve this [1]:

a)  Large reductions in losses in the energy sector by significantly increasing the efficiency of thermal power stations (combined heat and power, combined-cycle generation).

b)  $CO_2$ capture and sequestration in coal- and oil-fired power stations; with a significant limitation: the technology is not yet mature, is probably expensive and is not yet fully tested for environmental sustainability.

c)  Use of natural-gas power stations, substitution of coal and oil with natural gas. $CO_2$ emissions, compared to coal, are reduced to about 55% (to about 75% compared to oil), with a limitation: world-wide gas reserves are limited.

d)  Use of nuclear energy: the power stations do not emit $CO_2$; limitation: reserves of uranium, used in 3rd-generation reactors, are also limited. Use of reactors of the fourth generation is possible, but requires much technical and political deliberation. Nuclear fusion can only be considered in the second half of the century.

e) Use of all opportunities to generate electricity from water power; limitation: the potential for this is limited.

f) Use of wind energy: the technology is mature and, where the wind conditions are favourable, cost-efficient. With the use of off-shore installations, the potential is considerable.

g) Use of geothermal energy: Limitations: geothermal power stations are only suitable for locations where there are geothermal anomalies, and such locations have a large potential.

h) Use of biomass. Limitations: the capability of biomass is limited. Biomass should therefore be used primarily, and provided its use is, in ecological terms, acceptable, for fuel and heat, with the exception of local combined heat and power.

i) Use of solar thermal energy and photovoltaics. Solar thermal power stations are suitable for countries with a low proportion of diffuse light, and there they have a large potential. Photovoltaic generation is currently hindered by its high cost, but as its potential is practically unlimited and directly usable, its further development must be pursued with determination, including feed-in compensation (feed-in tariff).

# 3  Worldwide indicators for 2007

For this analysis, the world is partitioned as follows:

OECD-30  ---->    EU-15
                 USA
                 Japan
                 remainder OECD (13 countries)

non-OECD  ---->  China
                 India
                 remainder Asia/Oceania
                 Central and South America
                 Africa
                 Middle East
                 Transition countries (former Soviet Union +
                         non-OECD Europe)

In addition, our neighbouring EU-27 countries will first be considered, which are partly the EU-15, and partly belong to the other OECD countries (4 countries) and to the transition countries (non-OECD Europe, 8 countries).

The data gathered, shown and commented are:

- GDP (PPP) per capita,
- Energy intensity,
- $CO_2$ intensity,
- $CO_2$ indicator of sustainability

(the basis for this is the publications of the IEA of 2006 to 2009 and [9], see references).

## 3.1    European Union EU-27

**Population and gross domestic product**

The EU of 27 nations had, in 2007, nearly half a billion inhabitants (Fig. 1). Taking account of purchasing power parity (PPP), this generates a GDP of nearly 14800 billion year-2007 dollars, which is about 22% of world GDP (PPP). The 6 countries with the greatest populations (Germany, France, United Kingdom, Italy, Spain and Poland), with around 70% of the population of the EU, account for 73% of the GDP. The 15 countries of the EU-15 that are members of the OECD, with 79% of the population, account for around 88% of the GDP of the EU-27.

**Population of the EU-27**
2007,  total  496 million

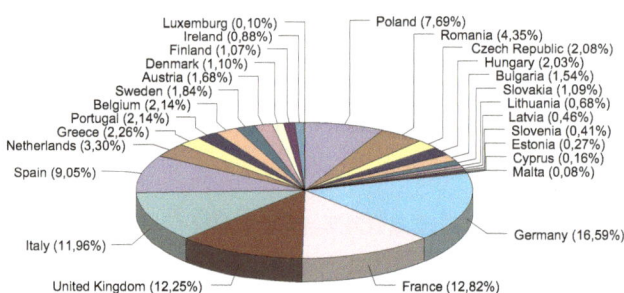

**Figure 1.**   Per cent partition of the population of the EU-27

**GDP/capita  (PPP)  in 10000 $/a**
EU, 2007

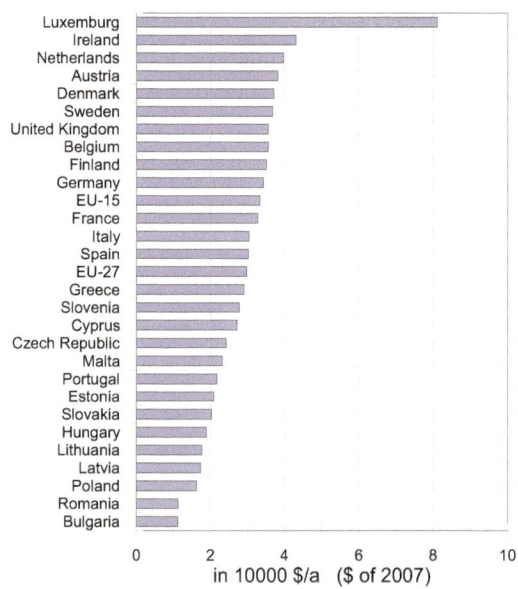

**Figure 2.**   Per cent partition of GDP/capita of the EU-27

The per-capita GDP (PPP) (Fig. 2) has an average of $30 000/a and varies (excluding the special case of Luxemburg) between $43 700/a in Ireland and $11 100 /a in Bulgaria. With the exception of Poland ($16 400/a), the 6 big countries are all in the range 30 000 to 36 000 $/a.

**Energy intensity (figure 3)**

The energy intensity is a measure of the energy required for producing the GDP (PPP). The mean for the EU is 1.6 kWa/$10 000  (gross domestic consumption). Speaking figuratively, the average required to generate $10 of purchasing power is the energy of about 1.6 litres of petrol. The group of EU-15 countries, at 1.5 kWa/$10 000  is slightly more efficient. The energy intensity figure is also influenced by the climate. The more southerly countries (such as Italy, Greece, Spain and Portugal) have a lower average energy intensity than the Nordic countries (such as Finland, Estonia and Sweden). Differences in energy efficiency are considerable. Ireland, for example, needs less than 1.1 kWa/$10 000, whereas Bulgaria requires 3.1 kWa/$10 000.

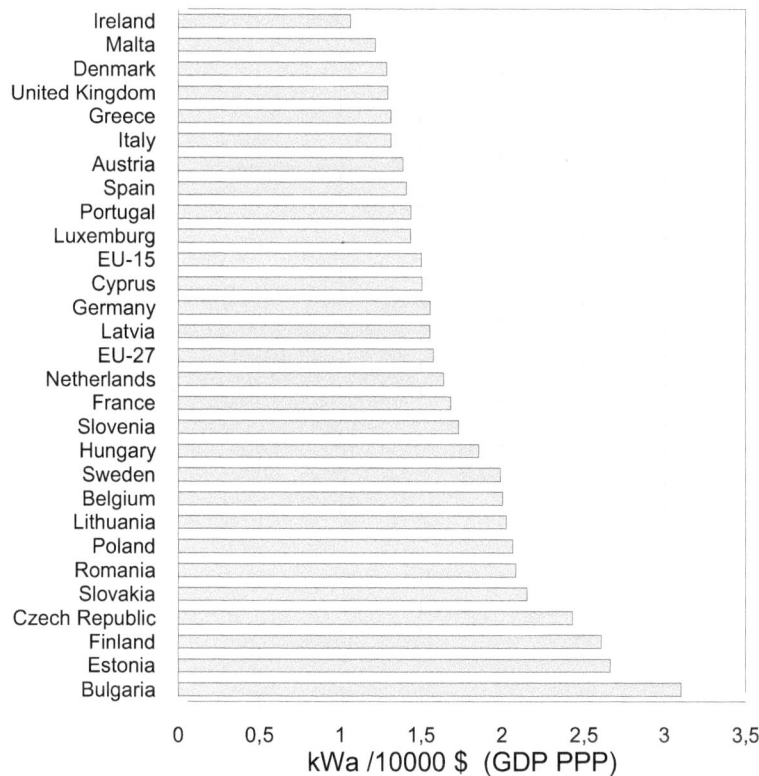

Figure 3   Energy intensity of EU-27 countries

## $CO_2$-intensity (figure 4)

Energy is only harmful to the climate when the associated $CO_2$ intensity is large. The mean value for this in the EU is 1.68 t $CO_2$/kWa. In the EU-15 it is slightly lower, at 1.63 t $CO_2$/kWa. Differences from this are large, with figures varying from 0.69 t $CO_2$/kWa in Sweden to 2.36 t $CO_2$/kWa in Poland. The way that electricity is generated is a very important factor. Countries where electricity is generated by water power and nuclear energy (Sweden, France) come off well in this respect, and those where the power stations are fuelled predominantly by coal, oil and gas (Poland, United Kingdom, Italy, Germany, and Spain) are worse than average.

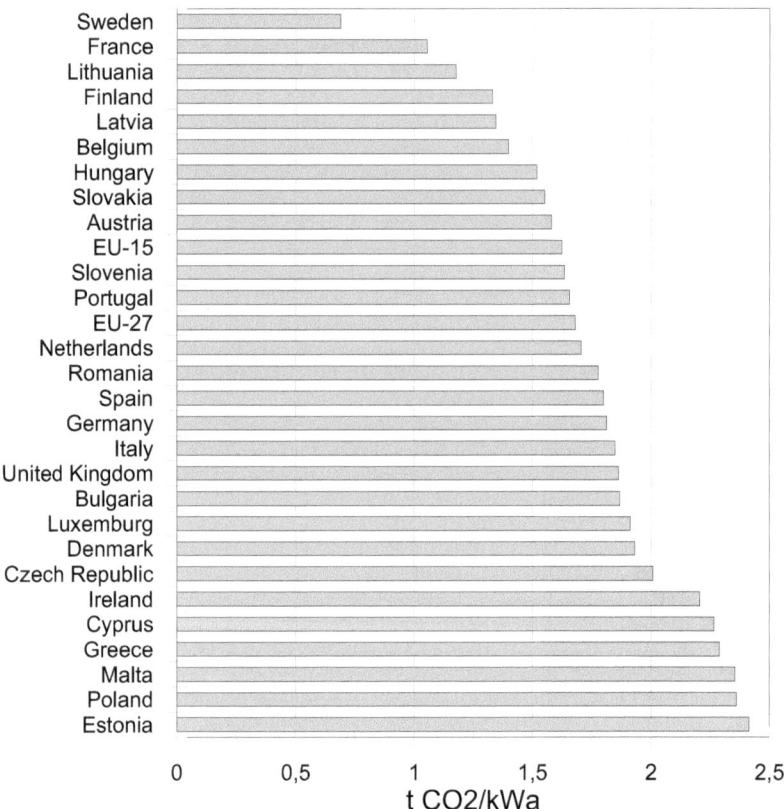

Figure 4  $CO_2$-intensity of the EU-27 countries

**Sustainability (figure 5)**

To judge the sustainability of the energy economy in climate terms, both factors are significant: energy intensity and $CO_2$ intensity. The product of these two quantities gives the $CO_2$ indicator in t $CO_2/\$10000$ or, more conveniently, in g $CO_2/\$$. The value for EU-27 is 265 g $CO_2/\$$ and that for EU-15 is 245 g $CO_2/\$$. Sweden and France are significantly below 200 g $CO_2/\$$, Italy and the United Kingdom slightly below 250 g $CO_2/\$$. Germany's energy economy has a poorer sustainability rating with 285 g $CO_2/\$$ and the figures for some east European countries are very bad, such as over 480 g $CO_2/\$$ for Poland and the Czech Republic, 580 g $CO_2/\$$ for Bulgaria and over 600 g $CO_2/\$$ for Estonia. Since climate has a clear influence (energy required for heating), a climate bonus or penalty should be introduced.

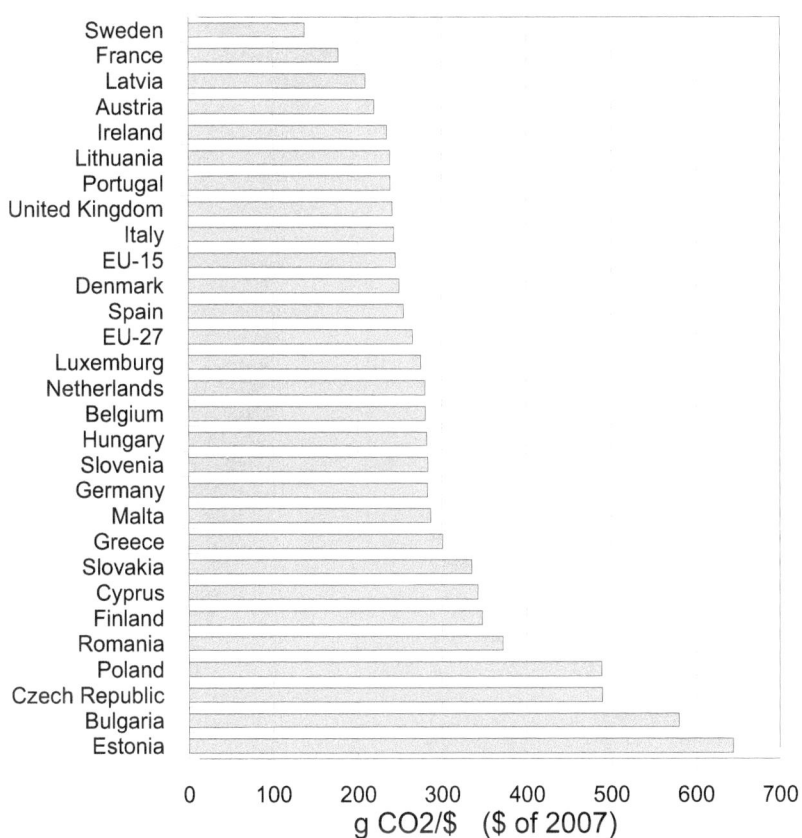

**Figure 5** $CO_2$ sustainability indicator of EU-27 countries

## 3.2  OECD- 30

### Population and gross domestic product

The 30 countries of the Organization for Economic Cooperation and Development (OECD) had a population of 1185 million in 2007, which is about 18% of the world's population (Fig. 6). They generate a GDP (PPP) of around $39000 bn, which represents 59% of the world GDP, after correction for purchasing power. The main players are the EU-15, the USA and Japan, which together have nearly 70% of the population and produce close to 80% of the GDP of the OECD.

**Population of OECD-30**
2007,  total  1185 million

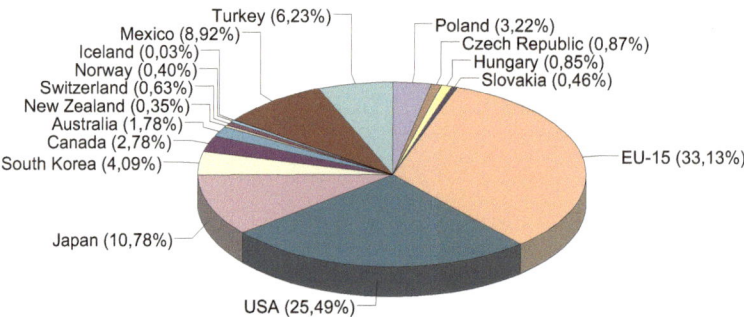

**Figure 6.**   Per cent partition of the population of OECD-30

**GDP/capita (PPP)  in 10'000 $/a**
OECD-30, 2007

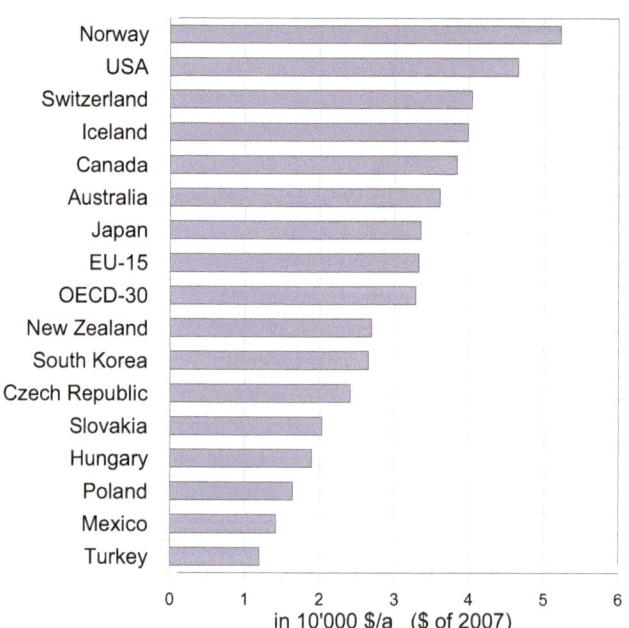

**Figure 7**   GDP (PPP) of the OECD-30 countries

The per-capita GDP (PPP) (Fig. 7) has an average of $32900 and varies between $52000 in Norway and $12000 in Turkey. After Norway, the leaders are the USA, Iceland and Switzerland. In addition to Turkey, the purchasing power is also relatively low in Mexico and some of the east European countries (Poland, Slovakia, Hungary), all between $12000 and $21000 per capita and year.

### Energy intensity (figure 8)

The average energy intensity (of the gross domestic consumption) in the OECD countries is 1.9 kWa/$10000, which is somewhat higher than in the EU-15. The very good figure for Switzerland (1.1 kWa) and the good figures for Norway, Japan, Turkey and the EU-15 are offset by the worse efficiencies of the USA (2.2 kWa), South Korea (2.3 kWa) and Canada (2.8 kWa). The very high figure for Iceland (> 5 kWa/$10000) is a consequence of the climate, but also of the great availability of $CO_2$-free energy sources in the form of water power and geothermal power.

**Energy intensity, kWa/10'000 $**
OECD-30, 2007

**Figure 8.** Energy intensity of the OECD-30 countries

## CO₂ intensity (figure 9)

Iceland is the leader with a very low value of $CO_2$ intensity (0.36 t $CO_2$/kWa), thanks to water power and geothermal energy. Norway and Switzerland also do well in the international comparison (slightly over 1 t $CO_2$/kWa) on account of their electricity generation purely from water power in one case and mainly water power and nuclear energy in the other. The mean value of $CO_2$ intensity of the OECD in 2007 was 1.78 t $CO_2$/kWa. The USA value is worse (1.86 t $CO_2$/kWa) and the worst is for Poland and Australia (> 2.3 t $CO_2$/kWa). The main reason for this is the high proportion of coal-fired power stations in the electricity production of these countries.

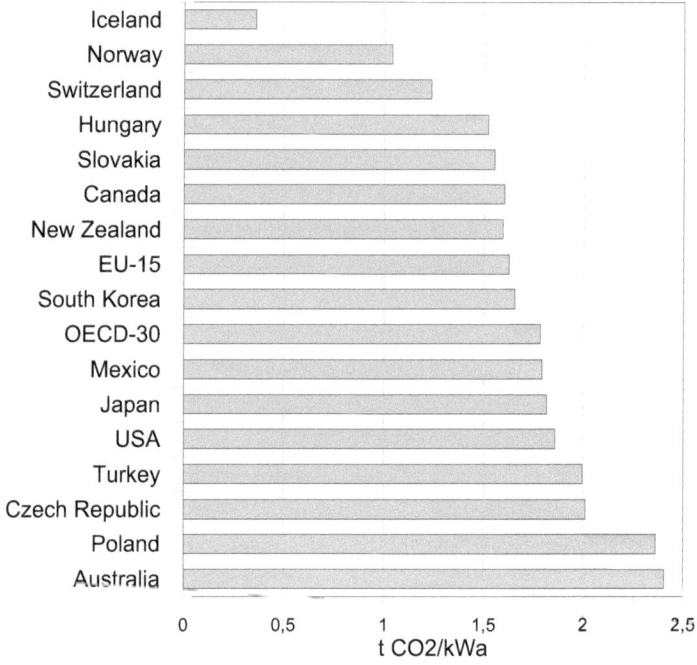

**Figure 9.** $CO_2$ intensity of the OECD countries

**Sustainability (figure 10)**

The sustainability of the energy economy in terms of climate change can be characterised by the product of energy intensity and $CO_2$ intensity. The $CO_2$ indicator calculated this way is 333 g $CO_2$/$ for the OECD. Values below 200 g $CO_2$/$ apply to Switzerland (leading with 139 g $CO_2$/$), Norway, Iceland and countries of the EU-15 such as Sweden and France. The reasons for this have already been mentioned (electricity wholly or mainly from hydroelectricity, nuclear or geothermal energy). At the other end of the scale are countries that generate electricity mainly from coal: USA, (410 g $CO_2$/$) and Poland, the Czech Republic and Australia (all nearly 500 g $CO_2$/$) or have bad energy efficiency (Canada 450 g $CO_2$/$, partly due to climate).

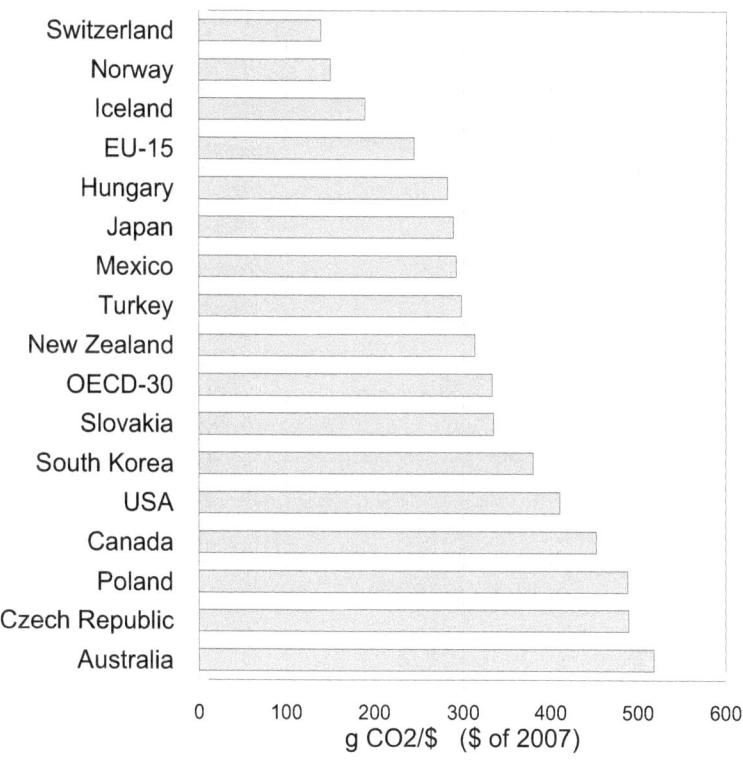

**CO2 emissions indicator, g CO2/$**
OECD-30, 2007

**Figure 10.** $CO_2$ sustainability indicator of the OECD countries

## 3.3   World

### Population and gross domestic product

For this analysis, the world is divided into the following zones: China, India, Rest-Asia/Oceania without OECD members, Central + South America, Africa, OECD (divided into EU-15, USA, Japan and the other 13 OECD countries), Middle East, former Soviet Union and non-OECD Europe. The population percentages are shown in Figure 11. The GDP (PPP) of the world is US $66 600 bn/a. However, there is a wide gap, with a factor of 5 between the OECD countries and the rest of the world. In 2007, with 18% of the world's population, the first group generated around 58% of the world GDP (PPP), while the second group, with 82% of the population, had to be satisfied with the remaining 42%. China, with 20% of the population, has a GDP (PPP) of $7 400 bn/a or 11% of the world GDP. India, with 17% of the population, has a GDP (PPP) of $3000 bn/a or 4.5% of the world GDP.

**World population**
2007,   total  6609 million

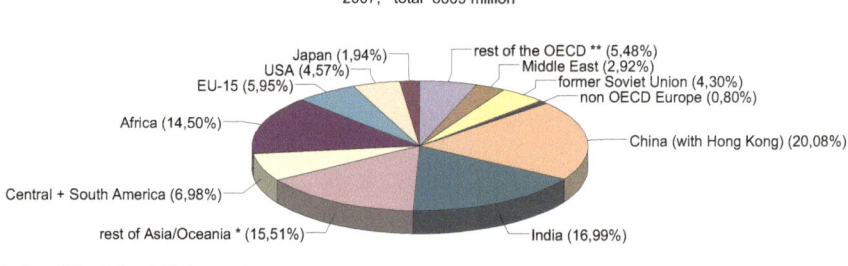

Japan (1,94%)
USA (4,57%)
EU-15 (5,95%)
Africa (14,50%)
Central + South America (6,98%)
rest of Asia/Oceania * (15,51%)
rest of the OECD ** (5,48%)
Middle East (2,92%)
former Soviet Union (4,30%)
non OECD Europe (0,80%)
China (with Hong Kong) (20,08%)
India (16,99%)

\* without China, India and OECD-countries
\*\* OECD without USA, Japan and EU-15

**Figure 11.**   Percent partition of the world population

**GDP/capita  (PPP)  in 10000 $/a**
World,  2007

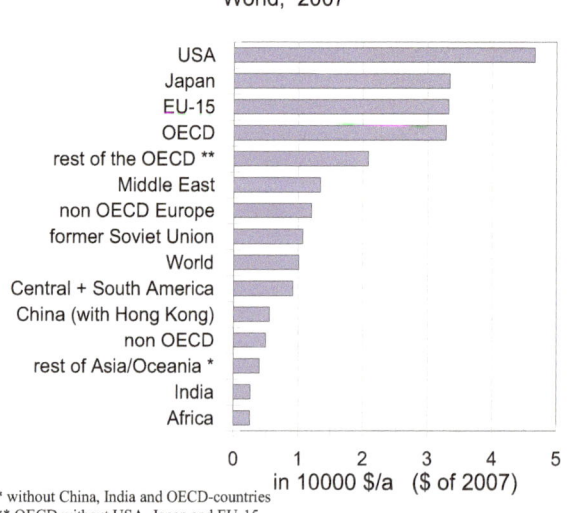

USA
Japan
EU-15
OECD
rest of the OECD **
Middle East
non OECD Europe
former Soviet Union
World
Central + South America
China (with Hong Kong)
non OECD
rest of Asia/Oceania *
India
Africa

0      1      2      3      4      5
in 10000 $/a   ($ of 2007)

\* without China, India and OECD-countries
\*\* OECD without USA, Japan and EU-15

**Figure 12.**   GDP (PPP)  of world zones and countries

The world average GDP (PPP) per capita (figure 12) is $10000 and varies between $30000/a and $50000/a in the industrialised world, and between $2500/a and $15000/a in the developing and emerging countries. The per-capita GDP (PPP) of China is $5600/a and that of India is $2700/a.

### Energy intensity (figure 13)

The world's mean gross energy intensity is 2.4 kWa/$10000. Clear differences can be observed all the way through the OECD and non-OECD countries. For example, the efficiency of energy use in the USA is only somewhat better than the world average, while the EU-15 and Japan have an energy intensity below 1.6 kWa/$10000. On the other hand, central and south America has an energy intensity comparable with that of the EU-15 or Japan. This could be partly due to underdevelopment, but also to the favourable climate. Energy wastage in the Middle East (nearly 3 kWa/$10000), in China (3.5 kWa/$10000) and in the former Soviet Union (4.4 kWa/$10000) is particularly blatant, which may be partly due to the low-cost availability of the fossil energy carriers.

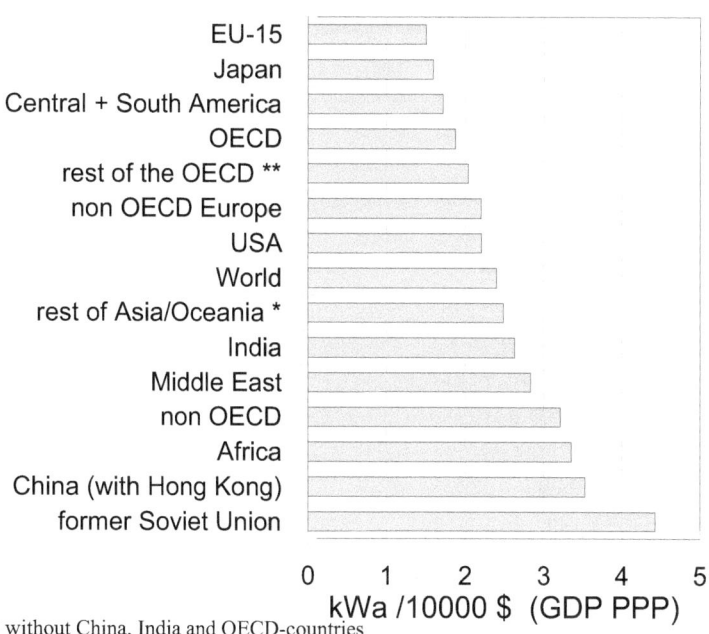

**Energy intensity, kWa/10000 $**
World, 2007

* without China, India and OECD-countries
** OECD without USA, Japan and EU-15

**Figure 13.** Energy intensity of the world zones and countries

## CO₂ intensity (figure 14)

The world-wide $CO_2$ intensity is 1.8 t $CO_2$/kWa, which is significantly higher than in the most advanced OECD countries. Exceptions are Africa (about 1 t $CO_2$/kWa) with an energy economy still partly based on biomass, and central and south America (1.44 t $CO_2$/kWa) thanks to the high proportion of hydropower in its electricity production. The former Soviet Union is slightly below the world average because of its energy economy that is largely based on natural gas. Countries above the world average are the USA, non-OECD Europe, the Middle East and, above all, China (the latter at about 2.3 t $CO_2$/kWa). Again, the main reason is the generation of electricity mainly from oil or coal.

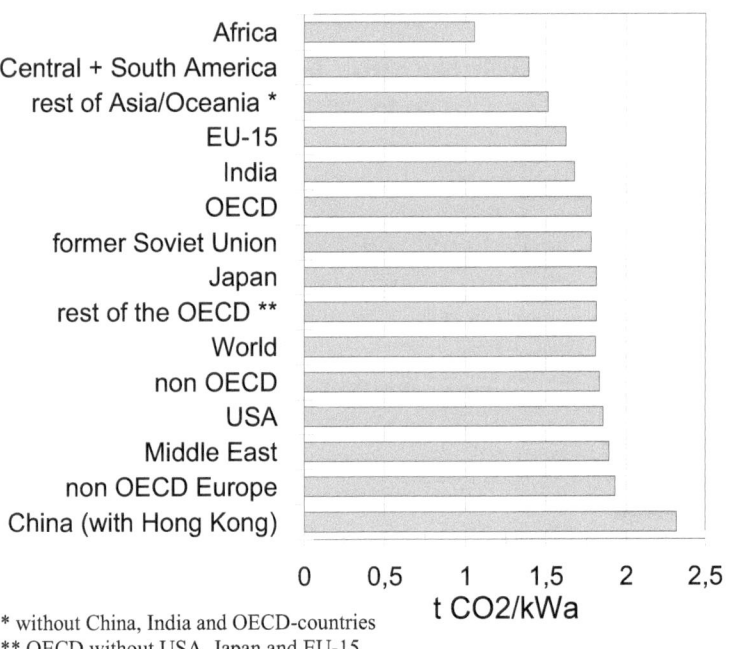

## CO2 intensity,  t CO2/kWa
### World,   2007

* without China, India and OECD-countries
** OECD without USA, Japan and EU-15

**Figure 14.**  $CO_2$ intensity of  world zones and countries

**Sustainability (figure 15)**

In connection with climate protection, the measure for sustainability of energy production and consumption is the $CO_2$ indicator obtained as the product of energy intensity and $CO_2$ intensity. The worldwide average of the $CO_2$ indicator is 435 g $CO_2$/\$. The countries below 300 g $CO_2$/\$ are central and south America, the EU-15 and Japan. With 410 g $CO_2$/\$, the USA has less sustainability. The greatest need for improvement is in non-OECD Europe countries (425 g $CO_2$/\$), the Middle East (535 g $CO_2$/\$), the former Soviet Union (nearly 800 g $CO_2$/\$) and China (more than 800 g $CO_2$/\$).

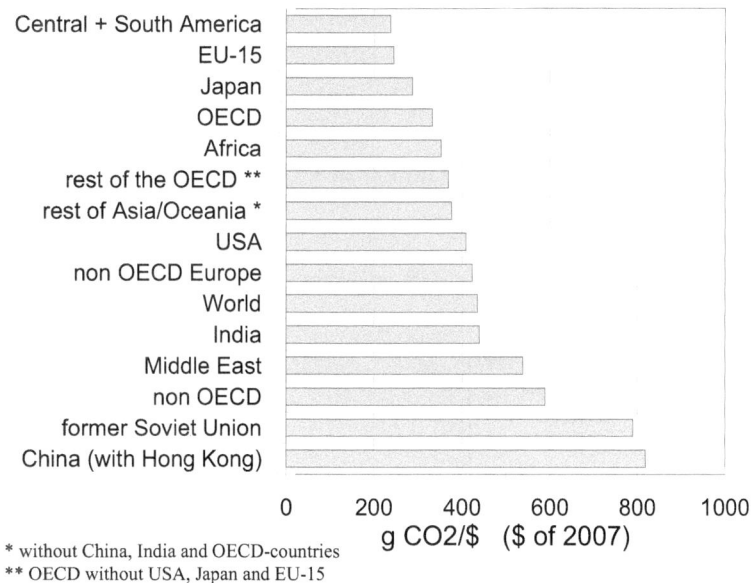

**CO2 emissions indicator, g CO2/\$**
World, 2007

* without China, India and OECD-countries
** OECD without USA, Japan and EU-15

g CO2/\$ (\$ of 2007)

**Figure 15.** $CO_2$ sustainability of world zones and countries

## 3.4    Rest of Asia/Oceania

**Population and gross domestic product**

The rest of Asia/Oceania is understood as the continents of Asia and Oceania without China, India, Japan and all further member states of the OECD such as South Korea, Australia and New Zealand. The total population is 1025 M (Fig. 16) and the GDP (PPP) close to $4200 bn/a ($ of 2007), or about 6.5% of the world GDP.

The 6 most populous states are Indonesia, Pakistan, Bangladesh, the Philippines, Vietnam and Thailand. With 76% of the population, they produce 60% of the GDP (PPP).

**Population of the rest of Asia/Oceania**
2007,  total  1025 million

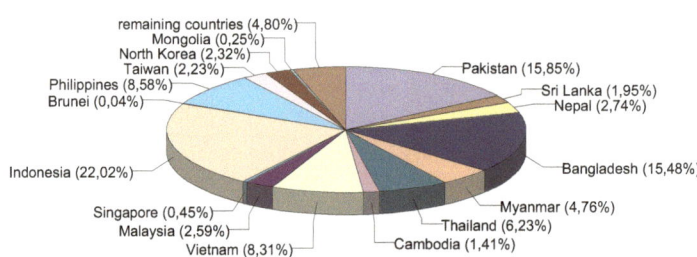

**Figure 16.**   Percent partition of the population of the rest of Asia/Oceania

## GDP/capita  (PPP)  in 10000 $/a
### rest of Asia/Oceania,  2007

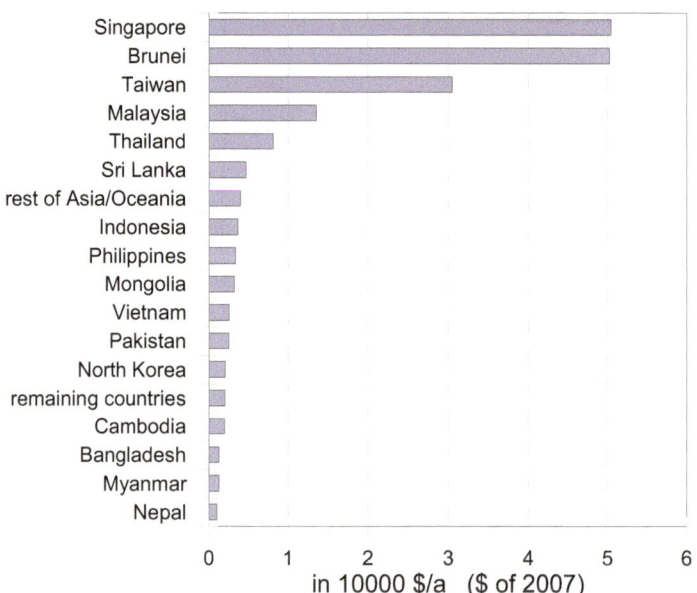

**Figure 17.**   GDP (PPP) of the rest of Asia/Oceania

The per-capita GDP (PPP) (Fig. 17) has a mean value of $4100/a, with that of the 6 most populous countries varying between 1300 in Bangladesh and 8200 in Thailand. OECD levels are attained only in Singapore, Brunei and Taiwan, with a GDP (PPP) between $30 000/a and $50 000/a.

**Energy intensity (figure 18)**

Thanks to the underdevelopment of many countries, the gross energy intensity, at 2.5 kWa/$10000, is only slightly over the world average. The energy economies of Indonesia, Thailand, Malaysia and Pakistan have low efficiency, between 2.6 and 3 kWa/$10 000. Extremely inefficient, and comparable with the countries of the former Soviet Union, are the energy sectors of North Korea (in spite the high proportion of hydroelectricity) and Mongolia (nearly 5 kWa/$10 000). EU-level has only Singapore with 1.5 kWa/$10 000. Taiwan, at around 2.1 kWa/$10 000 is significantly above Japan or the EU-15.

# Energy intensity,  kWa/10000 $
## rest of Asia/Oceania,  2007

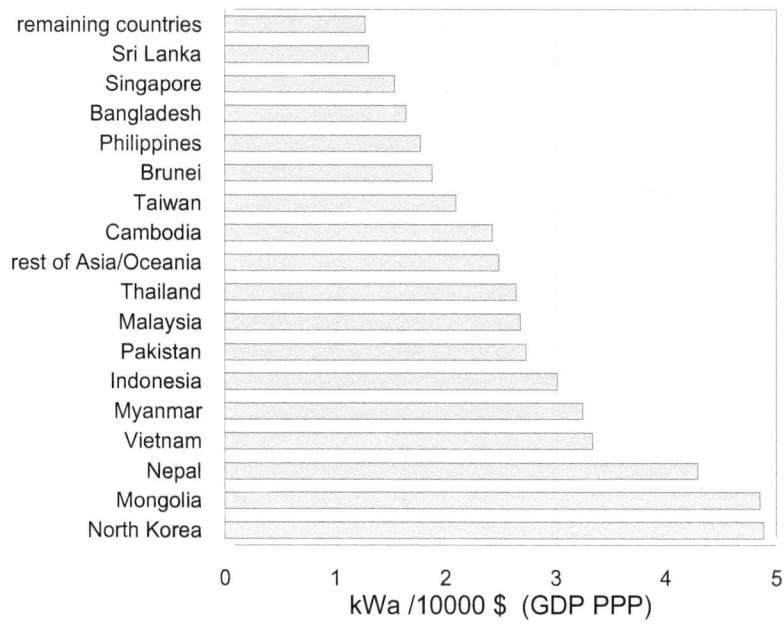

**Figure 18.**  Energy intensity of the rest of Asia/Oceania

**CO₂ intensity (figure 19)**

The average $CO_2$ intensity, at 1.45 t $CO_2$/kWa, is below the world average. Most of the countries with a low level of development have still lower values, with the exception of North Korea and Mongolia, whose energy economies, evidently dominated by coal, have the world's highest $CO_2$ intensity at over 2.5 t $CO_2$/kWa. Among the developed countries, measured by economic performance, the specific $CO_2$ emissions of Singapore are significantly lower than those of Taiwan.

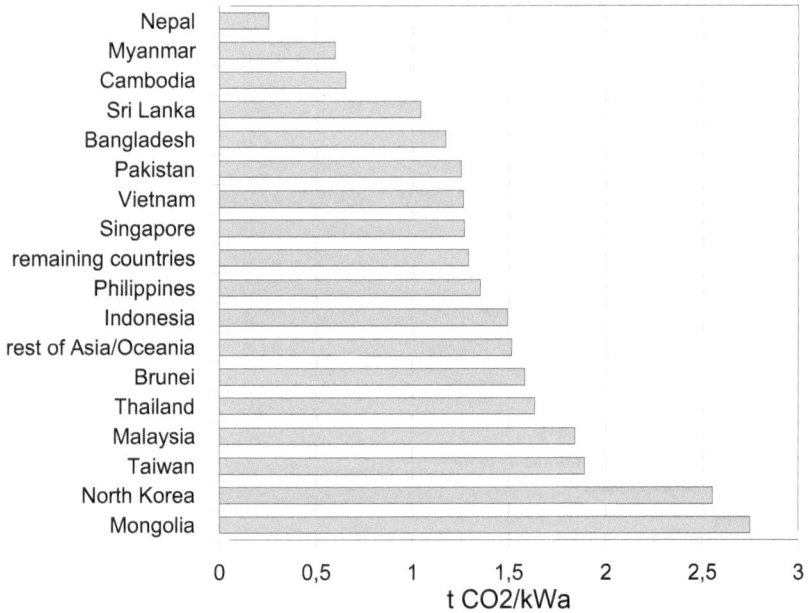

## CO2 intensity,  t CO2/kWa
### rest of Asia/Oceania,  2007

**Figure 19.**   $CO_2$ intensity of the rest of Asia/Oceania

## Sustainability (figure 20)

The $CO_2$ indicator for the rest of Asia/Oceania, with an average of around 375 g $CO_2$/\$ is significantly lower than the world average (435 g $CO_2$/\$) but above the OECD average (333 g $CO_2$/\$). The differences between the individual countries are very large. Among the populous and therefore more significant countries, Sri Lanka and Bangladesh, corresponding to their primitive energy sector with a large biomass component, and the Philippines, partly thanks to its water and geothermal power, have values below or slightly over 200 g $CO_2$/\$. Vietnam, Pakistan and Indonesia are in the OECD- or world-range. Singapore, which is economically developed, at around 200 g $CO_2$/\$, is better than EU-15. Taiwan at nearly 400 g $CO_2$/\$, has significantly less sustainability. Far removed from any sustainability are the energy economies of North Korea and Mongolia with more than 1200 g $CO_2$/\$.

# CO2 emissions indicator,  g CO2/$
## rest of Asia/Oceania,  2007

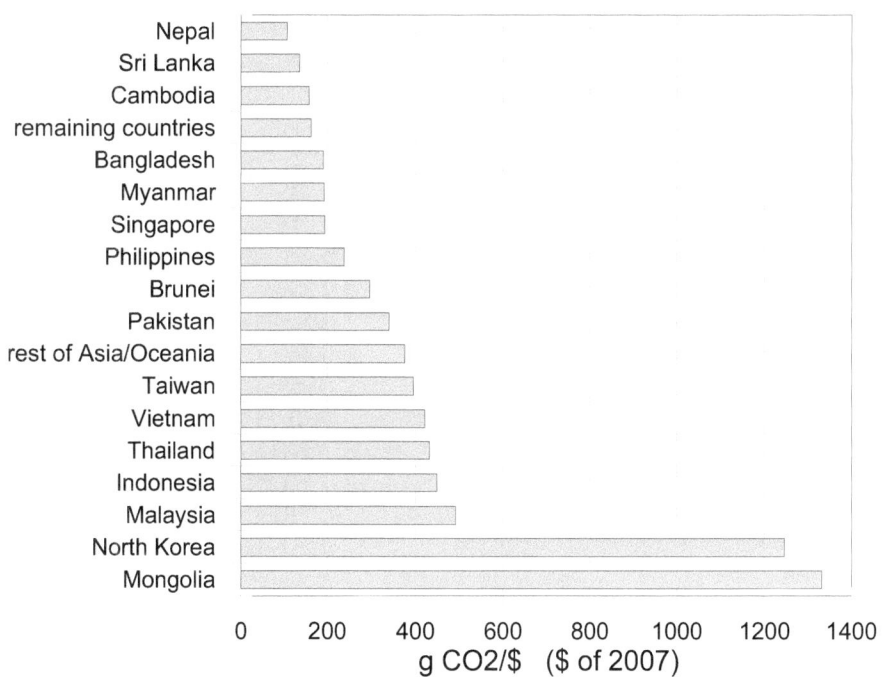

**Figure 20.**   $CO_2$ sustainability indicator of the rest of Asia/Oceania

## 3.5   Central and South America

### Population and gross domestic product

Central and South America, with 461 million inhabitants, have about 93% of the population of the EU-27, but with a GDP (PPP) that is 30% as large. Brazil, with 42% of the population and 42% of the GDP (PPP) is the most significant country of the continent (Fig. 21). The 6 most populous countries (Brazil, Columbia, Argentina, Peru, Venezuela and Chile) have over 75% of the population and produce 83% of the GDP (PPP).

**Population of Central + South America**
2007,   total   461 million

Dominican Republic (2,16%)
Jamaica (0,59%)
Netherlands Antilles (0,04%)
Trinidad and Tobago (0,29%)
remaining countries (0,88%)
Venezuela (6,09%)
Colombia (10,22%)
Ecuador (2,96%)
Peru (6,18%)
Bolivia (2,11%)
Chile (3,68%)
Paraguay (1,36%)
Uruguay (0,74%)
Argentina (8,75%)
Cuba (2,49%)
Guatemala (2,96%)
Nicaragua (1,24%)
Honduras (1,57%)
El Salvador (1,52%)
Costa Rica (0,99%)
Panama (0,74%)
Brazil (42,45%)

**Figure 21.**   Percent partition of the population of Central and South America

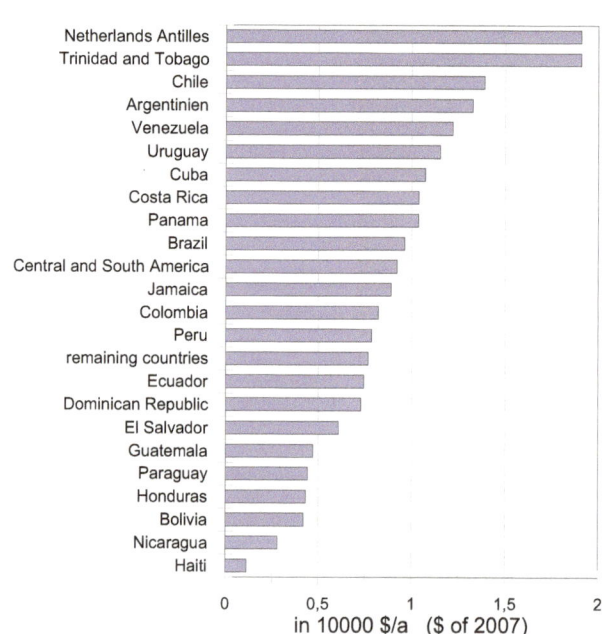

**GDP/capita  (PPP)  in 10000 $/a**
Central and South America, 2007

**Figure 22.**   GDP (PPP) of Central and South America

The per-capita GDP (PPP) (Fig. 22) has an average value of 9300 $/a. Brazil is slightly over the average. Among the larger countries, Argentina ($13 300/a), Chile ($13 900/a) and Venezuela ($12 200/a) are above average, and Columbia and Peru below average.

### Energy intensity (figure 23)

The overall energy intensity is 1.7 kWa/$10 000, which is about 71% of the world average and comparable with that of the EU-15. With a development stage of about 92% of the world average, the energy efficiency can therefore be assessed as relatively good. Of the 6 most populous countries, 3 (Peru, Columbia and Brazil) are below, and 3 (Chile, Argentina and Venezuela) are above average, in their energy intensity.

## Energy intensity,  kWa/10000 $
### Central and South America, 2007

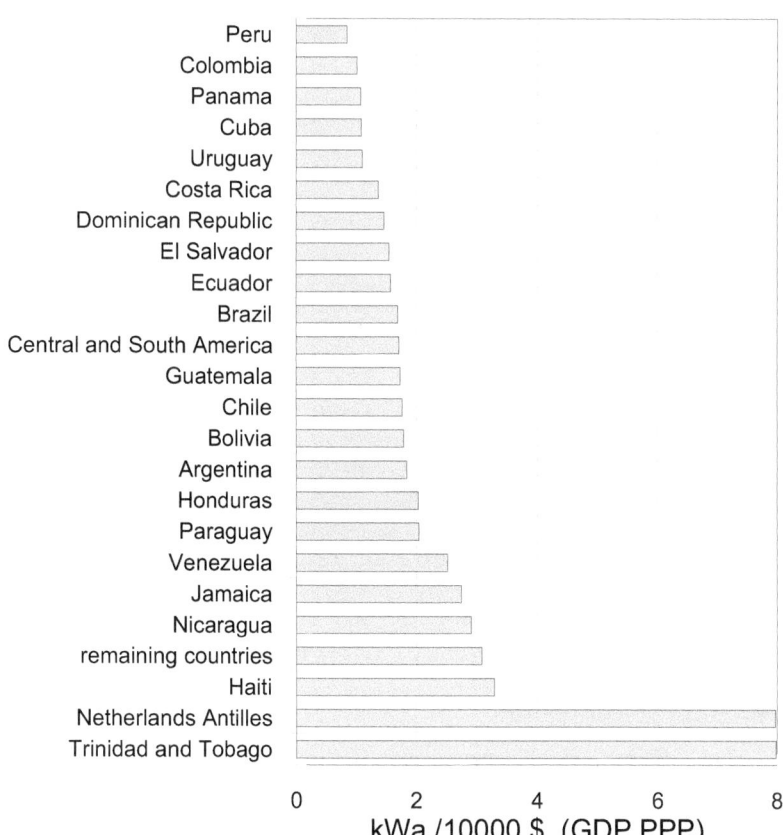

**Figure 23.**  Energy intensity of Central and South America

## $CO_2$ intensity (figure 24)

The $CO_2$ intensity is around 1.4 t $CO_2$/kWa, which is significantly below both the world average and that of the EU-15. This is partly attributable to the significant role of water power in electricity generation. The $CO_2$ intensity in Brazil is below average (1.1 t $CO_2$/kWa), partly thanks to the use of bio-fuels. The highest (among the 6 most populous countries) is the $CO_2$ intensity of Chile with 1.74 t $CO_2$/kWa,

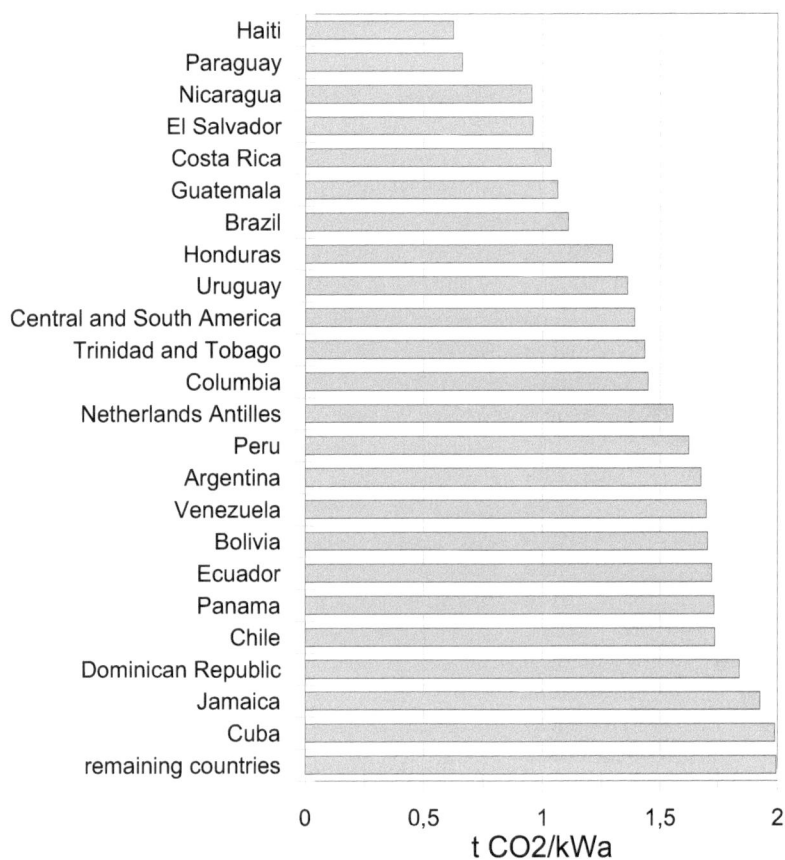

**CO2 intensity,  t CO2/kWa**

Central and South America,  2007

**Figure 24.** $CO_2$ intensity of Central and South America

**Sustainability (figure 25)**

The sustainability of the energy economy in South and Central America at 240 g $CO_2$/\$ is very good, which corresponds to the moderate intensities of energy and $CO_2$. This value even puts Central and South America in the lead (see Fig. 15). Three of the 6 large countries (Peru, Columbia and Brazil) are below this value. Venezuela is clearly out of line at 430 g $CO_2$/\$.

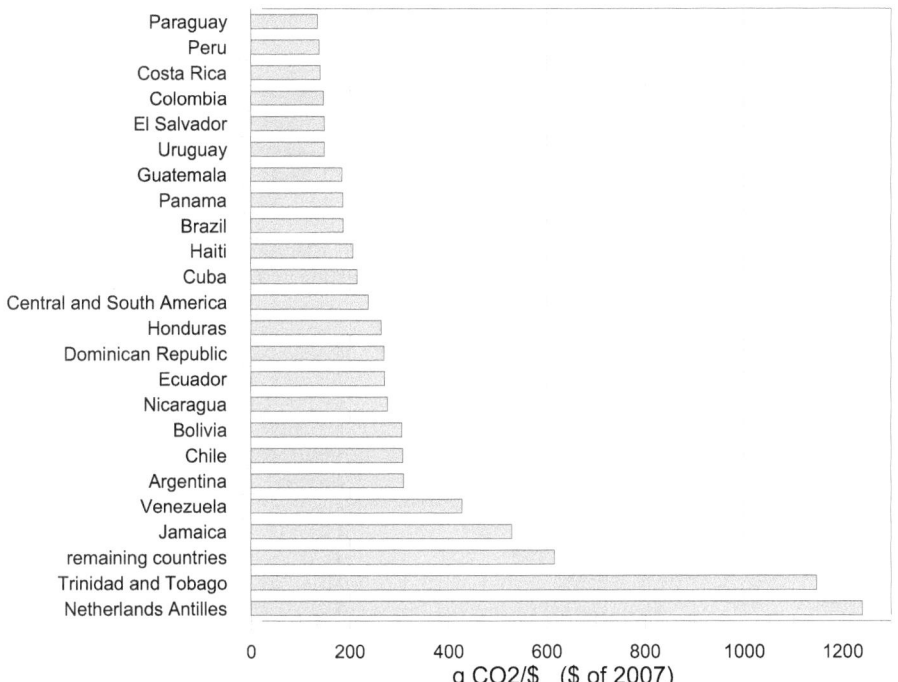

**Figure 25.**   $CO_2$ sustainability indicator of Central and South America

## 3.6   Africa

### Population and gross domestic product

With a population of 958 M (Fig. 26), Africa has a GDP (PPP) of $2500 bn. The five countries with the largest GDP (over 120 bn), which are South Africa, Egypt, Algeria, Nigeria and Morocco, together have 35% of the population and produce 61% of the GDP.

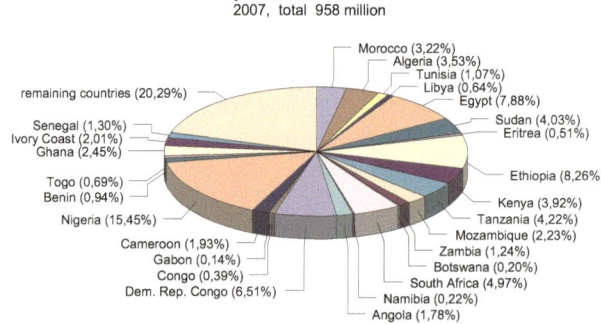

**Figure 26.**   Percent partition of the population of Africa

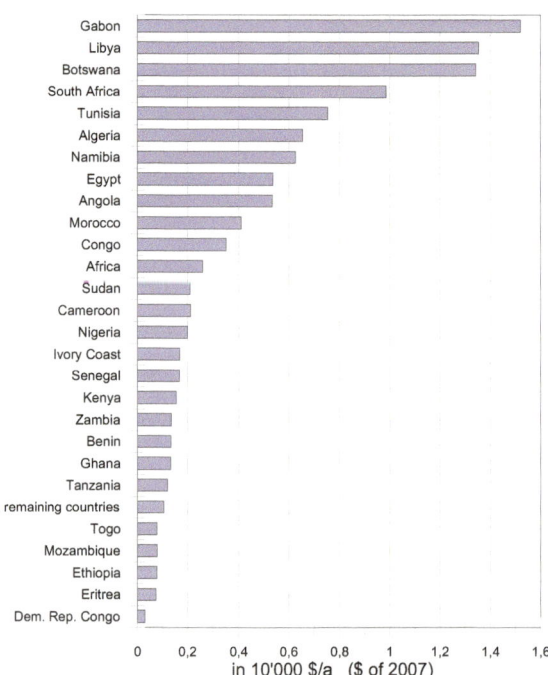

**Figure 27.** Percent partition of GDP/capita in African countries

The average per-capita GDP (PPP) (Fig. 27) is very low in the world comparison, amounting to $2600/a. Only four countries (Gabon, Botswana, Libya and South Africa) have or exceed $10 000/a. Six further countries (Tunisia, Namibia, Algeria, Angola, Morocco and Egypt) are in the range $4000/a to $8000/a. Most of the others are below the average for the continent, some by a large amount.

**Energy intensity (figure 28)**

The very high mean energy intensity of 3.35 kWa/$10 000, which is well over the world-wide mean and indicates an inefficient use of energy, is the consequence of underdevelopment but also of the relative wealth in fossil energy carriers that are made available at low cost (coal in South Africa, oil in Nigeria and Libya, gas in Algeria and Egypt etc.).

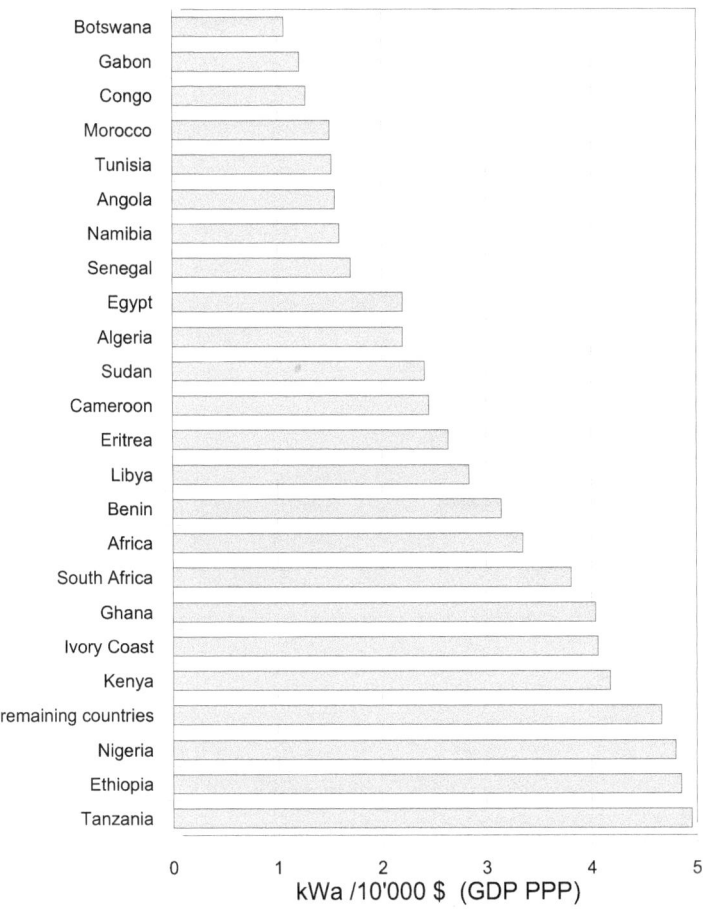

**Figure 28.** Energy intensity of the African countries

## CO$_2$ intensity (figure 29)

Africa has the world's lowest $CO_2$ intensity, at 1.05 t $CO_2$/kWa. The reason for that may be the continued high level of use of biomass that can be attributed to underdevelopment. On the other hand, the more developed countries have a $CO_2$ intensity that is close to the world average.

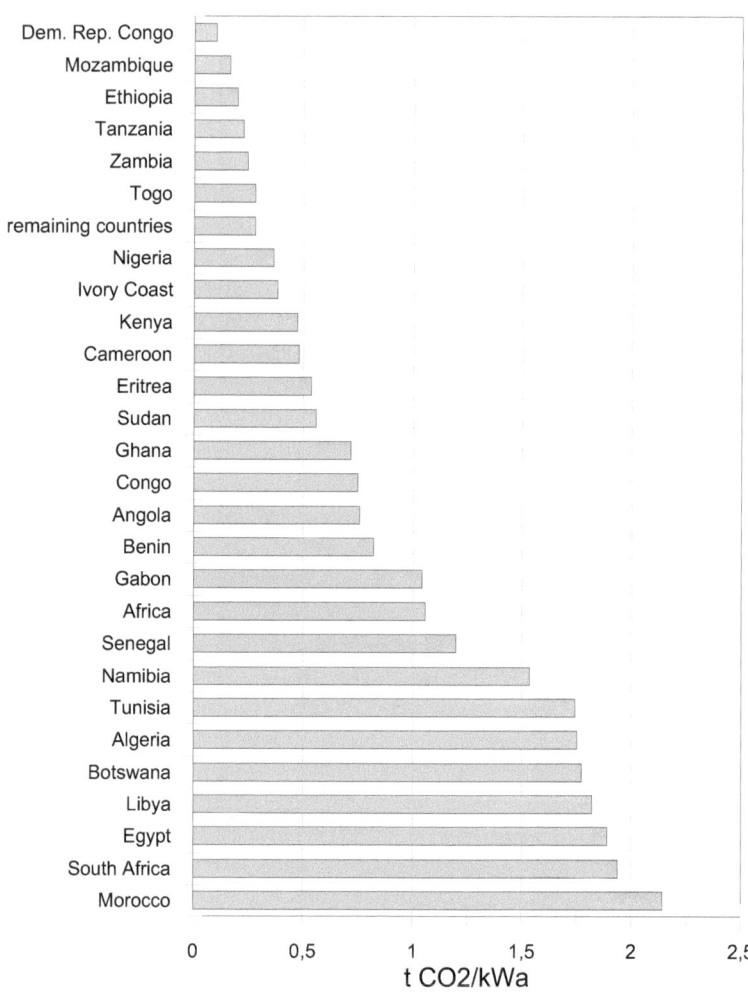

**Figure 29.** $CO_2$ intensity of the African countries

## Sustainability (figure 30)

The very low $CO_2$ intensity compensates the high energy intensity, so that the resulting $CO_2$ indicator of nearly 350 g $CO_2$/$ is not far from the worldwide middle ground. However there are large differences between the countries: from less than 150 g $CO_2$ in the very underdeveloped countries to nearly 400 g $CO_2$/$ and more than 700 g $CO_2$/$ (South Africa) in the countries with a strong oil and coal sector. Among the more developed countries, the leader Botswana (see Figs. 27 and 28 for per-capita GDP and energy efficiency) has less than 200 g $CO_2$/$, and Tunisia and Namibia, despite a relatively poor $CO_2$ intensity, have, in the world picture, a satisfactory $CO_2$ indicator of about 250 g $CO_2$/$.

## CO2 emissions indicator, g CO2/$
Africa, 2007

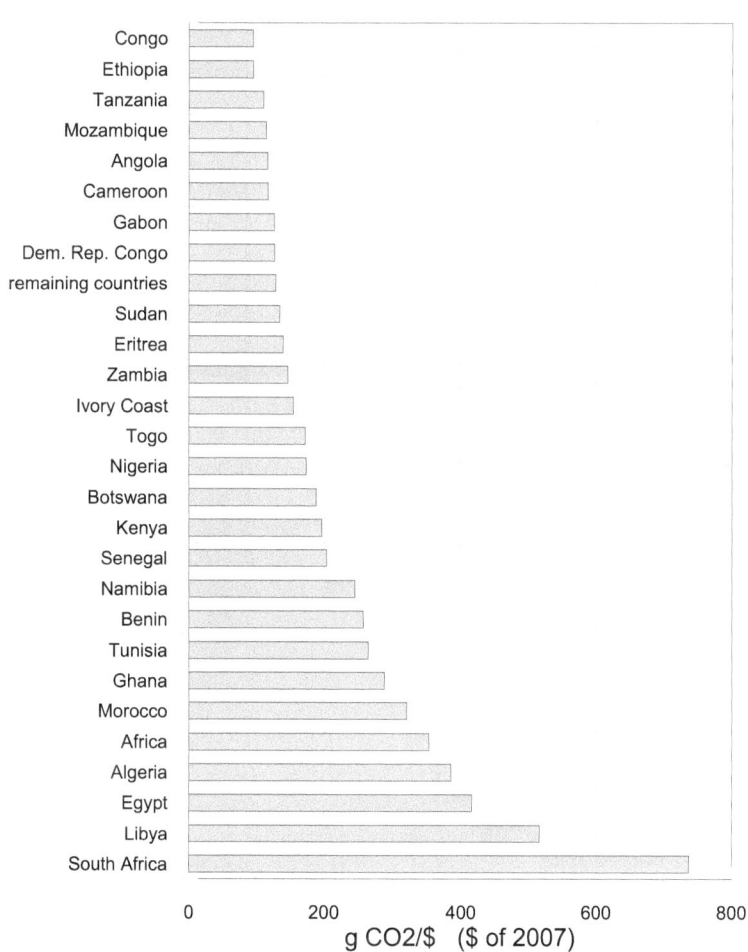

**Figure 30.** $CO_2$ sustainability indicator of the African countries

## 3.7    Middle East

### Population and gross domestic product

With a population of 193 M, the Middle East generates a GDP (PPP) of around
$2300 bn/a. About 53% of this is produced by Saudi Arabia, Israel and the Gulf
States (Oman to Kuwait), which together have 22% of the Middle East's population (Fig. 31). Iran, the state with the greatest population, generates a further 34%.
The remaining 13% is from the other states: Iraq, Syria, Yemen, Jordan and Lebanon, with a population share of around 41%.

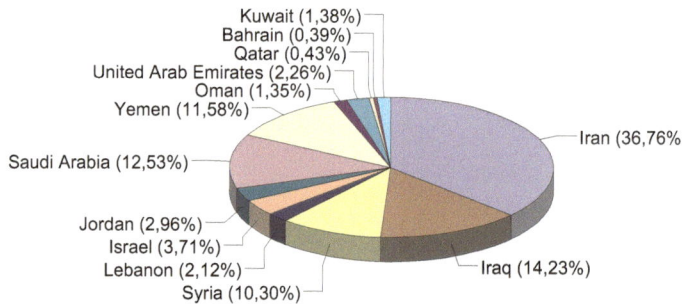

**Figure 31.**   Percent partition of the population of the Middle East

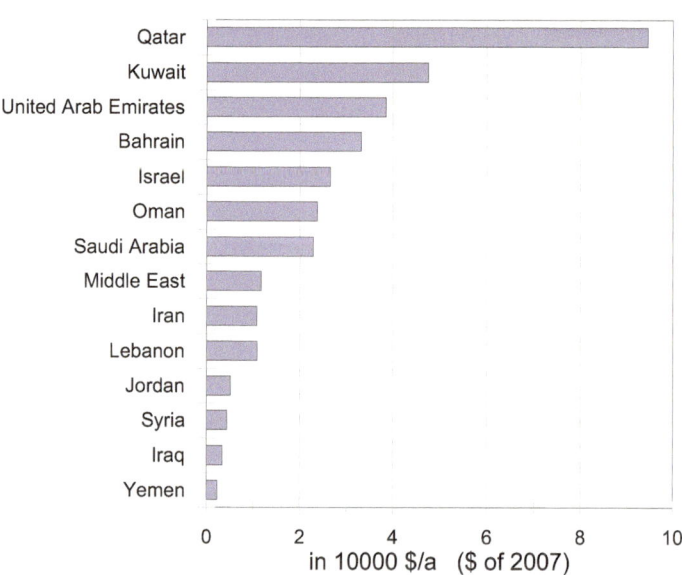

**Figure 32.**   GDP (PPP) of the countries of the Middle East

The difference in prosperity is correspondingly very large. The per-capita GDP (PPP) of the region (Fig. 32) is therefore, at $11800/a, somewhat over the world average. The most countries in the first list of names have a GDP (PPP) of $23000/a to $50000/a, which is similar to the level in the EU or the OECD. Exception is Qatar with more than $90000. In contrast, the purchasing power of the last-named countries is only in the range $200/a to $1100/a. The very low level for Iraq is partly due to the war.

**Energy intensity (figure 33)**

The mean energy intensity of about 3.2 kWa/$10000 is above the worldwide mean, indicating an considerable waste of energy. This applies particularly to Qatar, Saudi Arabia and the United Arab Emirates, with nearly 4 kWa/$10000. The exceptionally high value for Iraq must be judged with care because of the consequences of the war and the continuing disturbances. Lebanon and Israel are the clear leaders with an energy intensity below 2 kWa/$10000, an energy efficiency comparable with those of the EU-15.

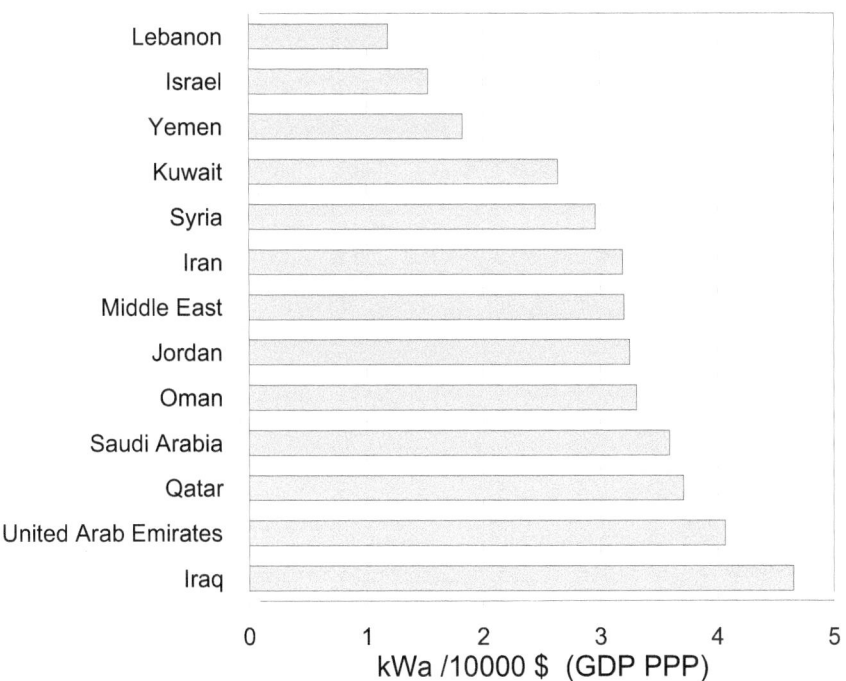

**Figure 33.** Energy intensity of the countries of the Middle

**CO₂ intensity (figure 34)**

The mean $CO_2$ intensity of 1.9 t $CO_2$/kWa is not significantly above the world-wide mean value of 1.8 t $CO_2$kWa. The differences from one country to another are not very large. Oman is the most advanced at 1.74 t $CO_2$/kWa, while Israel at 2.26 t $CO_2$/kWa is at the other end of the scale.

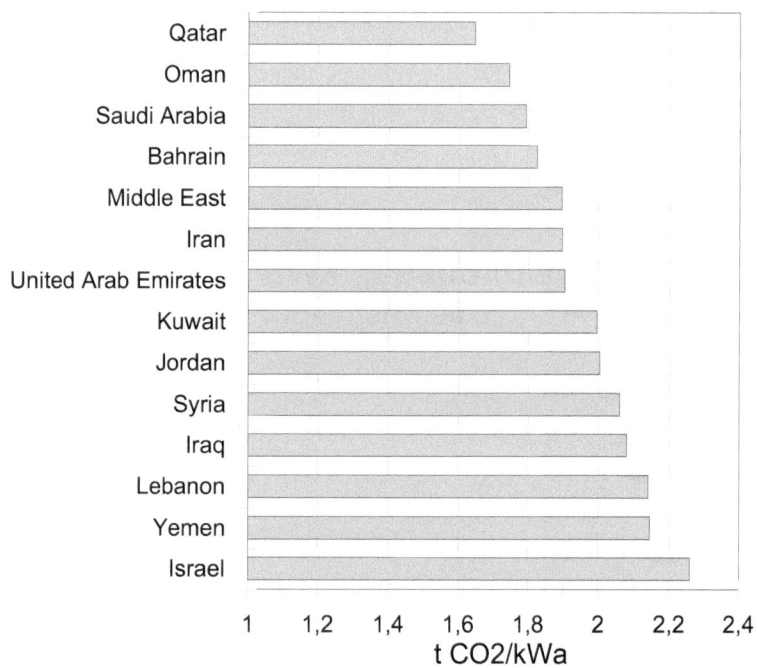

**Figure 34.** $CO_2$ intensity of the countries of the Middle East

**Sustainability (figure 35)**

With a mean value of 600 g $CO_2$/$, the energy economy of the Middle East is anything but sustainable. Especially in the Gulf States, but also in Saudi Arabia and Iran, a more reasonable use of energy should be aimed at in the medium-term, with a big increase in energy efficiency, despite their oil riches. Only Israel, with its good level of energy efficiency, and despite a poor $CO_2$ intensity, has a sustainability indicator comparable to those of the OECD countries.

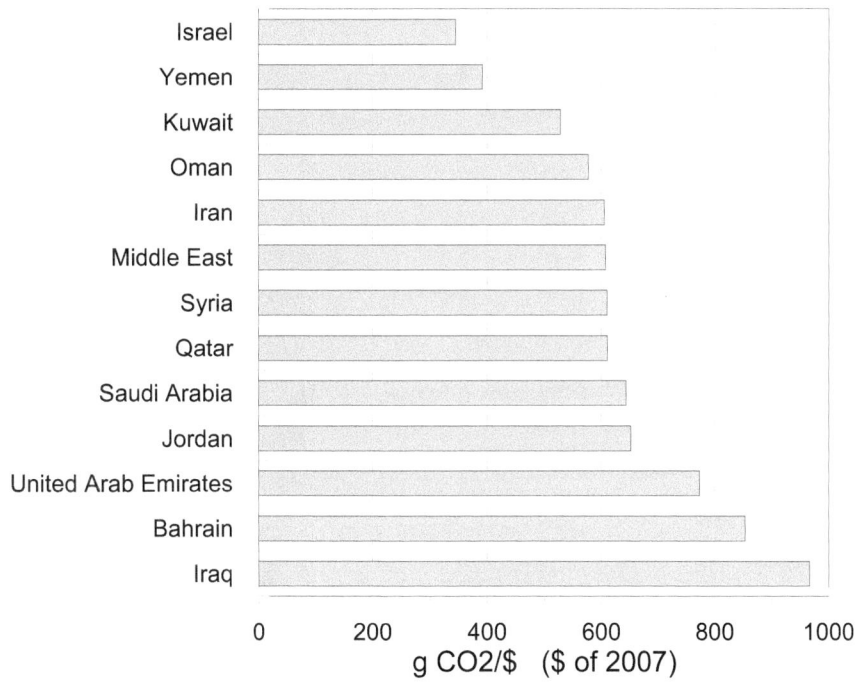

**Figure 35.** $CO_2$ sustainability of the countries of the Middle East

## 3.8    Transition countries

### Population and gross domestic product

The term "transition countries" is used to indicate the 15 countries of the former Soviet Union (with 84% of the population and 83% of the purchasing-power-corrected GDP (PPP)) and 11 European countries that are not members of the OECD (with 16% of the population and 17% of the GDP (PPP), Fig. 36).

In the former Soviet Union, Russia with 50% of the population generates about 69% of the GDP (PPP). Together with the Ukraine and Belarus, the population share is 70% and the share of GDP (PPP) is 83%. The Baltic States have now become members of the EU-27.

**Population of the transition countries**
2007,  total  337 million

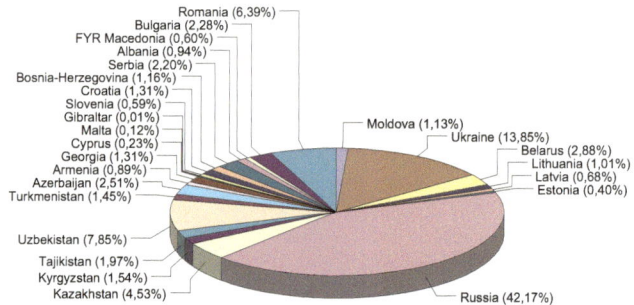

**Figure 36.**   Percentage partition of transition-country populations

**GDP/capita  (PPP)  in 10000 $/a**
Transition countries,  2007

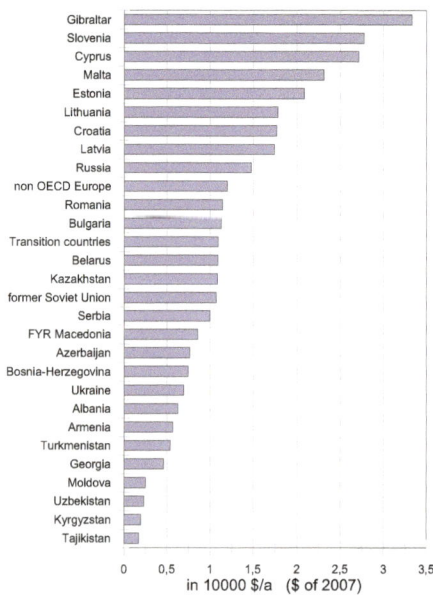

**Figure 37.**   GDP (PPP) of the transition countries

Of the 11 non-OECD countries, Romania and Bulgaria (which have also become members of the EU-27), having together 54% of the population and 52% of the GDP (PPP), are the most significant. Kosovo and Montenegro are not included in the IEA statistics

The average per-capita GDP (PPP) (Fig. 37) in the transition countries is $11000/a (in Russia $14800/a and around $18400/a in the Baltic States), with an average of $12000/a in the European non-OECD countries; Slovenia and Cyprus in particular are significantly higher at over $25000/a.

**Energy intensity (figure 38)**

Overall, the transition countries have an energy intensity of 4 kWa/$10000 which is significantly above the world-wide average and represents the world-wide greatest waste of energy. In the former Soviet Union the value is 4.4 kWa/$10000 (determined by Russia and Ukraine), but only 1.6 to 2.7 kWa/$10000 in the Baltic States. Uzbekistan has an exceptionally high value, at 10 kWa/$10000. In the European non-OECD countries the average, at around 2.2 kWa/$10000, is significantly lower, but still somewhat above the OECD average. Only Albania, Croatia, Slovenia and Cyprus are below 2 kWa/$10000.

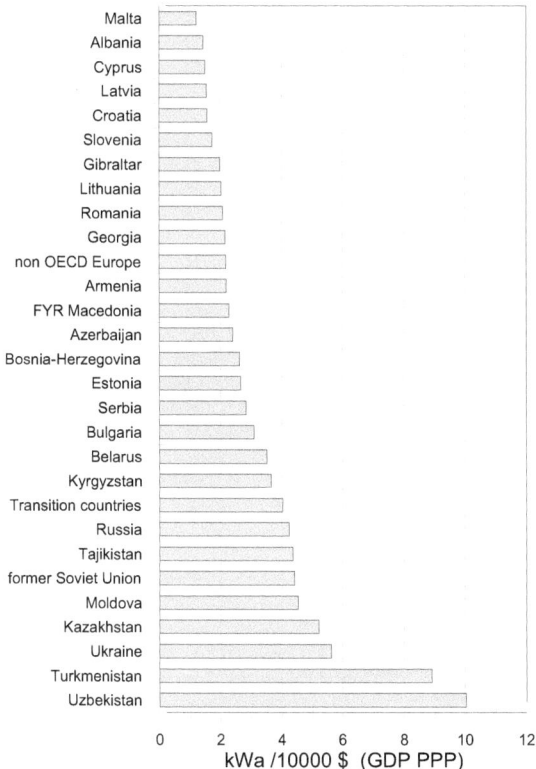

**Figure 38.** Energy intensity of the transition countries

## $CO_2$ intensity (figure 39)

The $CO_2$ intensity is not far from the world-wide average and in the European non-OECD countries it is somewhat higher than in the former Soviet Union (higher consumption of coal and less of natural gas). The $CO_2$ intensity is particularly high (significantly over 2 t $CO_2$/kWa) in Kazakhstan, FYR Macedonia, Serbia, Estonia and Bosnia-Herzegovina.

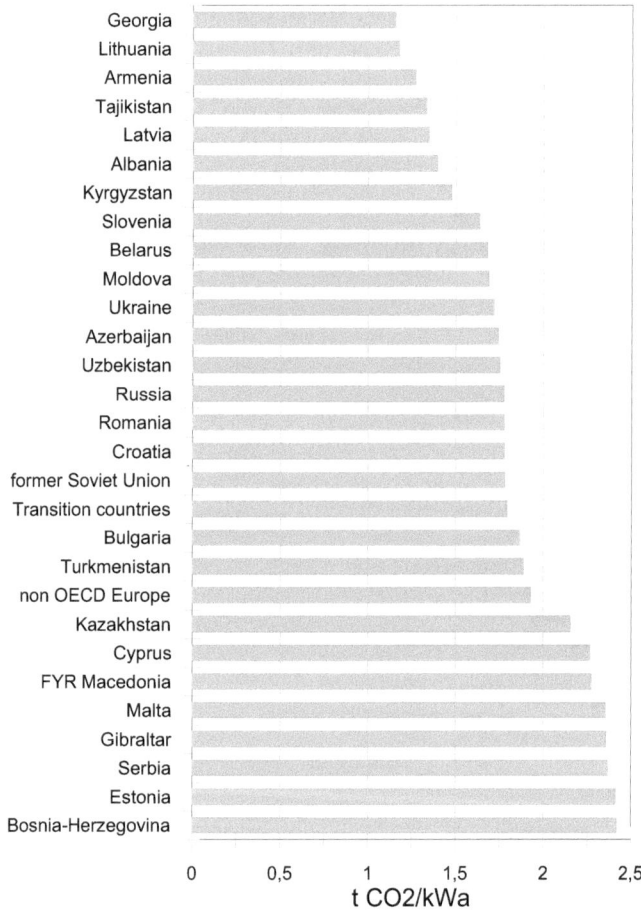

**CO2 intensity,  t CO2/kWa**
Transition countries,  2007

**Figure 39.**  $CO_2$ intensity of the Transition countries

**Sustainability (figure 40)**

In the European non-OECD countries, the $CO_2$ indicator has an average of 420 g $CO_2$/\$, and nearly 800 g $CO_2$ in the former Soviet Union. The efficiency of energy use is poor in nearly all the transition countries, with a negative effect on the sustainability of their energy sectors. Only Slovenia, Croatia and Armenia are below 300 and Georgia, Lithuania, Latvia, and Albania below 250 g $CO_2$/\$

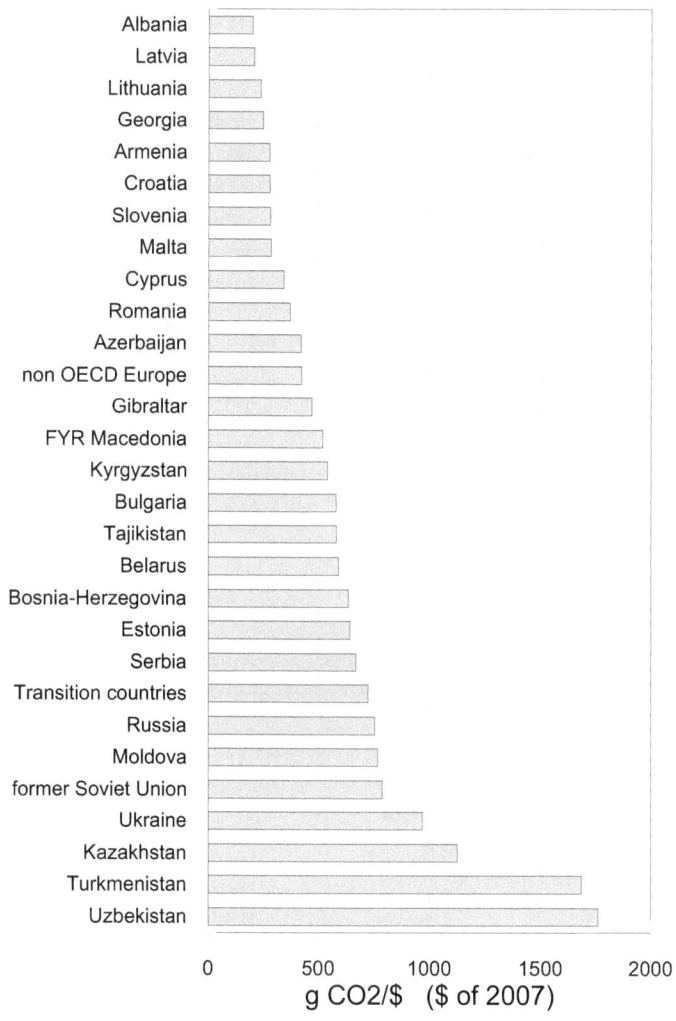

**Figure 40.** $CO_2$ sustainability indicator of the Transition countries

# 4  Indicator values required for 2030

To meet the stabilisation condition (Section 1), the world-wide emissions of $CO_2$ in the year 2030 must be below 27 000 Mt. With a GDP (PPP) of $135 000 bn ($ of 2007), this would give a necessary $CO_2$ sustainability indicator of $\eta = 2$ t $CO_2/\$10 000$ (or 200 g $CO_2/\$$). Switzerland, for example, has already gone below this value in 2007 (see Section 3.2, Fig. 9).

In the same scenario, the gross energy requirement would be 19.7 TWa (growth 1.1%/a). This corresponds to an energy intensity of $\varepsilon = 1.46$ kWa/$10 000. The formula $k = \eta/\varepsilon$ gives the mean $CO_2$ intensity $k = 1.37$ t $CO_2$/kWa required for climate protection. Comparing with the year-2007 value of 1.8 t $CO_2$/kWa (Section 3.3, Fig. 14) shows that $CO_2$ intensity must improve to 0.76 of that value. This improvement is not unrealistic and seems to be within reach if appropriate structural changes are expedited round the world in the energy sector, and above all in the generation of electricity (Section 2). The world-wide indicators for energy and $CO_2$ emissions would then be $e = 2.43$ kW per capita at that time (increase of about 3% over the 2004 figure) and $\alpha = 3.33$ t $CO_2$ per capita (reduction of 20% below 2004).

The following graphs show, for the OECD, USA, non-OECD and China, the annual changes in the indicators from 2004 to 2007 and the annual percent changes required until 2030 for protecting the climate.

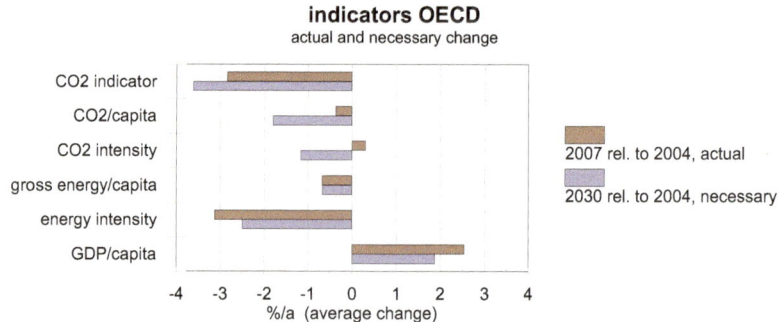

**Figure 41.**  OECD indicators, effective change 2004-2007 (red) and required change until 2030 for climate protection for the stated GDP (PPP) change (blue)

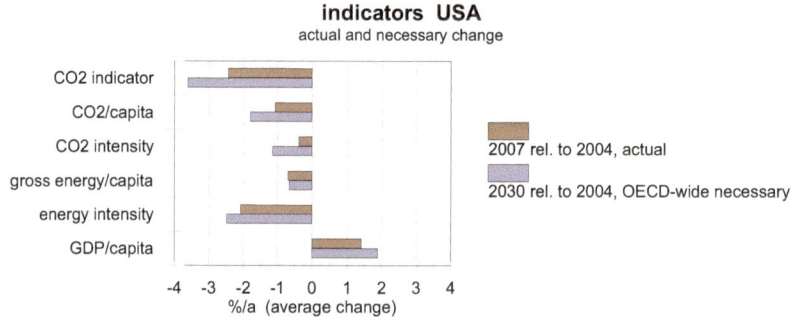

**Figure 42.**  USA indicators, effective change 2004-2007 (red) and required change in the OECD until 2030 for climate protection for the stated GDP (PPP) change (blue)

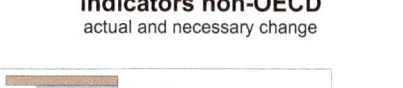

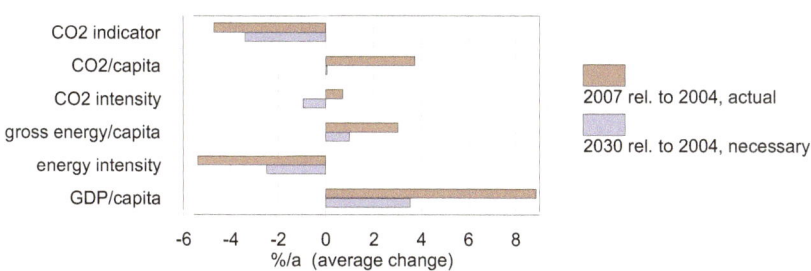

**Figure 43.** Indicators, non-OECD, effective change 2004-2007 (red) and required worldwide change by 2030 for climate protection with stated GDP (PPP) change (blue)

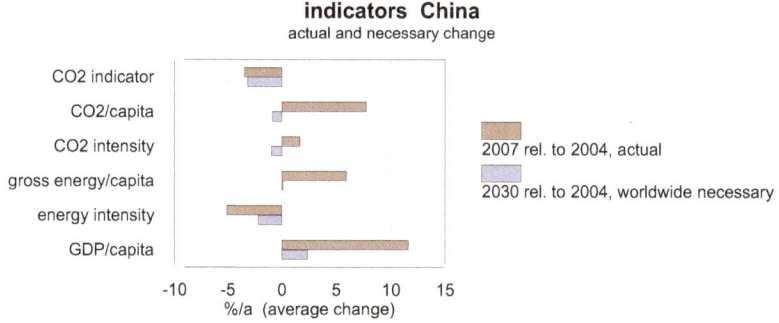

**Figure 42.** Indicators for China, effective change 2004-2007 (red) and worldwide required change until 2030 for climate protection with stated GDP (PPP) change (blue)

If the growth in GDP is greater than assumed, a correspondingly greater reduction of the $CO_2$ indicator is required. The corresponding changes for the world as a whole have already been given and commented in the introduction.

# 5   Indicator values required for 2050

The halving of $CO_2$ emissions in the next 20 years to 13 500 Mt $CO_2$ requires even greater efforts which, however, seem entirely possible, if the efficiency improvements and structural adjustments that have been successfully introduced up to 2030 are further pursued with determination. To estimate the indicator values, assume, for 2050, a world population of 9 bn and a GDP (PPP) of around $220 000 bn ($ of 2007), which corresponds to a world-wide growth of 2.5%/a from 2030. Assume that gross energy consumption still increases, but more slowly, to 22.5 TWa (growth 0.67%/a). The resulting worldwide indicators would then be:

- Energy intensity $\varepsilon \approx 1$ kWa/$10 000 $ (efficiency improvement by a further 30%),
- $CO_2$ intensity k $\approx 0.6$ t $CO_2$/kWa (further strong improvement to 0.44 of the previous value),
- Sustainability indicator $\eta \approx 60$ g $CO_2$/$.

The world-wide indicators for energy and $CO_2$ emissions would then be: $e \approx 2.5$ kW per capita (increase by nearly 3% compared to 2030) and $\alpha = 1.5$ t $CO_2$ per capita (large reduction by 55% relative to 2030). The new *Blue Scenario* and the *450-Scenario* of the IEA [8] are much closer to these demands.

## 6   Concluding remarks

The climate protection goals formulated above are only achievable with the united efforts of all the heavily populated and economically developed parts of the world. The USA, the EU, Japan and the other OECD members, China and India together had, in 2007, 55% of the world's population and generated 77% of the GDP (PPP). It is essential for these countries to work together and for this to include further transition or emerging countries with large demographic or energy potential such as Russia, Brazil and countries of the Middle East. *Energy intensity, $CO_2$ intensity* and *their product* must be criteria that are accepted everywhere, reliably determined and improved in terms of local purchasing power. In addition, emissions trading should be implemented more rigorously and the possibility of introducing a world-wide $CO_2$ tax should be examined. It would be an advantage if the available IEA statistics could be updated more quickly. The reliability of determining the gross domestic product, with purchasing power correction, is equally important.

The current crisis in finance and the economy does not make this exercise any easier. It can also be assumed that the data of many countries for GDP and energy consumption will change significantly as a consequence of the recession. It is not clear what effect this will have on the $CO_2$ results. We are entitled to await the data for 2008 to 2010 with some anxiety. They will show how seriously the world community takes the idea of climate protection.

## References:

[1]   Crastan V.: Elektrische Energieversorgung 2, second edition, Springer, 2009
[2]   IEA, International Energy Agency : World Energy Outlook, 2006
[3]   IEA: Key World Energy Statistics, 2009
[4]   IEA: Key World Energy Statistics, 2006
[5]   Stocker T.: Die Erde im Treibhaus, Bulletin SEV/VSE, Bern, 2007 (1)
[6]   IPCC (Intergovernmental Panels on Climate Change): Fourth Assessment Report, 2007
[7]   Missbrauchte Klimaschutzpolitik, Neue Zürcher Zeitung, 18 July 2009
[8]   IEA, International Energy Agency: Special early excerpt of World Energy Outlook, 2009
[9]   IMF World Economic Outlook Database, 2009

**Annex**

**Energy economics and climate change**

# A.1 Basic terms, historical review

Figure A.1 represents the structure of the supply of energy, showing the energy carriers in use today and the possible future energy carriers. We distinguish between four energy conversion stages: *Primary energy, secondary energy, final energy* and *useful energy.*

Companies with activities in the *extraction, conversion* and *transport* of energy carriers form the *energy sector of the economy.* Their task is to provide the *end-user* with energy in the form of the desired energy carrier (*final energy*) The consumer uses *energy-usage processes* to convert the final energy into useful energy.

## A.1.1 The energy sector

*Primary energy carriers* are sources of energy that occur naturally. For the most part, they are not used at the places where they occur, but are first extracted (e.g. mined), then transported and, if necessary, converted into another more suitable form of energy (secondary energy carrier). Coal and natural gas are usually simply extracted and transported to the place of use; oil, on the other hand, is converted into heating oil and motor fuel in refineries; naturally occurring uranium is converted into nuclear fuel; Coal, oil, natural gas, water power, wind power, nuclear fuels, solar heat and waste generate electricity in power stations.

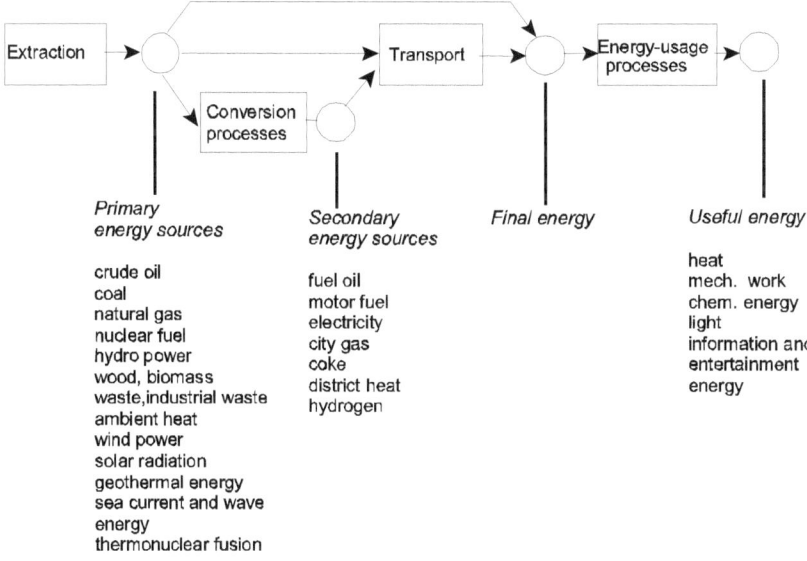

**Figure A.1**   Energy forms and conversion stages [A.5]

A few energy carriers, such as solar radiation and environmental heat, can be used in a decentralised way as heat or for generating electricity directly at the place where they arise.

In addition to the most important *secondary energy carriers* that have already been mentioned such as heating oil, motor fuel and electricity, Fig. 1 also shows hydrogen as a secondary energy carrier that could be important in the future (more details in [A.5], chapter 8).

There are costs, losses and environmental pollution associated with *extraction, conversion* and *transport.*

Under normal conditions and when dealt with correctly and competently, the processes of *mining* and *transport* have only a small adverse environmental impact, but in wartime situations and when there are accidents with fossil and nuclear fuels, they can have very serious environmental consequences. Some examples are: accidents with tankers, sabotage and fires in conveyor equipment and oil and gas pipelines, accidents in the transport of radioactive fuel elements.

In the *conversion* processes the harm is, above all, the constant environmental pollution from the *burning of fossil fuels* (combustion products, $CO_2$ that damages the climate). Further localised sources of environmental pollution can be accidents in nuclear power stations, oil storage tanks and refineries.

In addition, the *conversion* of thermal energy from fossil and nuclear fuels into mechanical or electrical energy always produces a high level of *waste heat* for thermodynamic reasons ([A.5], sections 5.1-5.3). Given suitable investments, it is possible to make use of some of this heat (combined heat and power, [A5], section 5.4).

*Final energy* or *final consumption* is the energy actually available to the consumer. *National energy statistics* usually refer to final consumption and/or primary energy or gross consumption (definition: Section A.3.2). The figures differ considerably because of the losses arising in conversion and transport, which includes the energy consumption of the energy sector itself. The *losses of conversion* are very high. *Transport losses* arise primarily in the transmission and distribution of electrical energy which, for example in Switzerland, has an overall efficiency of 93%, with the greatest part of the losses occurring in the distribution network.

## A.1.2   Energy-usage processes

Final energy is converted by the energy-usage processes of energy consumers (households, industry, services, businesses, agriculture, traffic) into *useful energy*, mainly *heat*, *mechanical work* and *light* (Fig. A.1). A small part is stored in the form of *chemical energy* in end products (steel, aluminium etc.). The amount of this is insignificant, at least in Switzerland. Another small part, but this is likely to increase in the future, is the energy used in *information technology* and *entertainment* (for computers, leisure electronics and communication).

*Energy-usage processes* have very different efficiencies. Whereas, for example, in an electrical heater the electricity is 100% converted into heat, on average only about 20% of the energy in the fuel of a present-day automobile is converted into mechanical energy and a conventional incandescent light bulb converts only about 5% of the electrical energy into light (but at least in winter, the rest of the energy is not lost but contributes to heating the buildings). The energy-usage processes are therefore in some cases associated with large losses of energy; the average efficiency has been estimated, for example for Switzerland in 1997, as 56% [A.20]. The emissions of $CO_2$ gases produced by burning heating oil and motor fuel are *locally* and *globally* a considerable source of *environmental pollution* (section A.7).

### A.1.3   Historical review

The *uses made* of final energy have changed only little and slowly since the beginnings of human history. Only the range of available energy carriers is much greater, and the technologies of energy conversion and usage have become more numerous, efficient and convenient.

**Mechanical work**

Today, as in previous times, *mechanical work* is used for acquiring and creating goods, for providing services and for transporting goods and people. The muscular efforts of humans and animals were prerequisites for survival in almost all early societies, both for the hunter-gatherers and even more so for the farmers. In ancient times, urban societies could only exist and develop thanks to slave workers. Over the course of time it became possible to make more effective use of muscle power and increase productivity through many mechanical inventions such as the wheel and the lever and, later, all kinds of mechanical devices, and also making use of the natural forces of water and wind (water-wheel, windmill, sailing ships), in other words to exploit new sources of energy. Much later, starting in the 18th century, was the beginning of replacing muscle power by more powerful steam machines. From the end of the 19th century, electric motors enabled a big increase in productivity and in the 20th century, internal combustion engines resulted in greatly increased mobility.

**Heat and light**

Today, as in earlier times, *heat* and *light* serve to protect from the cold, in the provision of food, increasing security, improving conditions of work, human well-being and thus the development of cultural activities. Energy sources were originally wood, plant waste and dried dung. Hot springs were already in use in ancient times. Coal first came gradually at the end of the 17th century and then mineral oil, natural gas and electricity which were not in widespread use for generating heat until the 20th century.

Heat, mostly at high temperature, was also used for producing various kinds of goods (metals, earthenware, jewellery and art objects). In this connection, we speak today of industrial and commercial *process heat* (in contrast to *heat for cooking and comfort*).

**Summary**

The *structure of making energy available and using it* changed but little for millennia until developments in science from the end of the 18th century led to the *age of technology*. The beginnings of this were already present in the ancient world and the middle ages. But only in recent times were revolutionary technical means invented for the mechanical use of the heat energy of fuels (steam engines, and later, combustion engines). Success was achieved in creating new (secondary) energy carriers such as town gas and electricity and in distributing them to small-scale consumers. *Electricity* in particular simplified and encouraged the use of energy on a scale previously unknown and, together with *coal*, made the industrial revolution possible. In many countries *water power* gained great significance in generating electricity. Coal, initially the most important primary energy carrier, was to an increasing extent replaced by *mineral oil* after the 2nd World War, i.e. in the 2nd half of the 20th century, but in many countries it retains a primary position in the generation of electricity. Finally, new primary energy sources were successfully exploited such as *natural gas* and *nuclear fission*.

The structure of the energy industry underwent a fundamental change in the course of a century. The muscle power of humans and animals is still used today (e.g. the bicycle as a means of transport) but this is omitted from the statistics of the energy industry as non-commercial energy. Wind and water power are only taken into account to the extent that they contribute to the production of electricity.

## A.1.4  Prospects and problems

The progress of civilisation in the 20th century would have been considerably slower without cheap energy. The *use of energy* in industrialised societies increased by a factor of more than ten in the course of that century. It freed humanity from the burden of heavy manual work and made a decisive contribution to a previously unimaginable mass prosperity. Although this process initially only affected a part of the world, it was the basis for a world-encompassing improvement in the material conditions of life.

An important element of this *progress* was firstly the mechanisation of manual labour, and later its replacement by automation. Connected to this are rapid changes in social structures and a progressive intellectualisation of work in general. The social problems that arise from this are a challenge for the socio-economic order, including in the developed world, and cannot be solved simply by an often inadequately thought-out globalisation. More of the dark sides of this development became manifest in recent decades with *overloading and poisoning of the biosphere* and *threats to the stability of the climate*.

Attempts are being made to rectify this through political demands for *social and ecological sustainability*. The contribution of the energy industry lies in the implementation of making energy available with the *least possible damage to the environment* and in *environmentally sustainable* usage of energy (sections A.6 and A.7). In this connection, the *rational use* of energy is very important, i.e. the improvement of the efficiency levels of all processes.

## A.2   Availability of primary energy

All the primary energy carriers listed in Figure A.1 can ultimately be traced back to the two principal forms that occur in the universe, *gravitational* and *nuclear energy*, as is shown in Figure A.2.

The primary energy carriers can be divided into two large classes that will be discussed below:

- *non-renewable energies:* fossil and nuclear energy carriers,
- *renewable energies:* tidal energy, geothermal energy and
  above all, direct and indirect solar energy.

Table A.1 shows the *energy content* values of important energy carriers. Since various units are used in the literature, conversion factors are indicated.

**Table A.1**   Energy content of energy carriers (mean values [A.3]) and conversion factors

| Oil | 10 000 kcal/kg | Conversion factors: |
|---|---|---|
| Black coal | 6 700 kcal/kg | |
| Lignite | 4 800 kcal/kg | 860 kcal/kWh |
| Wood | 3 600 kcal/kg | 4.19  kJ/kcal |
| Waste | 2'840 kcal/kg | 3.6  MJ/kWh |
| Natural gas | 8 660 kcal/m³ | 31.54  GJ/kWa |
| Gas from coal and | | 0.753 toe/kWa |
| oil (town gas) | 4 200 kcal/m³ | 1.12 tce/kWa |
| Uranium 235 | 20 Tcal/kg | 1 kWa = 8760 kWh |

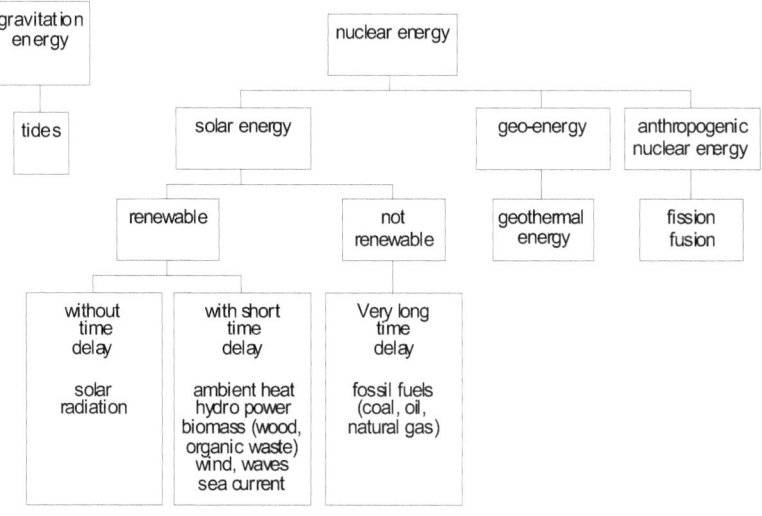

**Figure A.2**  Origins of available primary energy types

## A.2.1    Non-renewable energies

### A.2.1.1 Future demand

Coal, mineral oil, natural gas, fissionable materials (uranium, thorium) are not renewable energy sources. The fusion process has not yet been implemented ([A.5], chapter 9). In 2004, about 87% of primary energy requirements were met from non-renewable energy sources. According to the scenarios of the IEA (International Energy Agency) this proportion will decrease only slightly by 2030: to 86% according to the reference scenario and to 84% according to the alternative scenario (Fig. A.3; more details in section A.5).

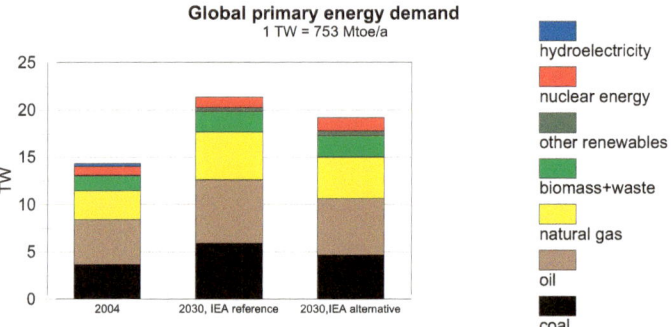

**Figure A.3**   World energy requirement (primary energy) in 2004 and 2030, in the IEA reference and alternative scenarios [A.10]

If we assume, for example, that the mean energy requirement increases linearly from 15 TW in 2004 to 25 TW in 2050, the cumulative energy requirement in this period would be 920 TWa. The question arises, concerning the finite reserves and resources of energy carriers, whether and how they are in a position to meet around 85% of this requirement, about 780 TWa. This quantity would, according to the IEA scenarios (Fig. 1.3) be made up as follows: coal: 230 TWa, mineral oil: 280 TWa, natural gas: 210 TWa, Uranium 60 TWa.

### A.2.1.2  Reserves and resources

*Reserves* are verifiable deposits that can be used at today's market prices and by means of today's technologies. *Resources* are deposits that are known to exist but at present cannot be economically extracted, or those that are assumed and "probably discovered in the future and/or used with technologies yet to be developed and sold at what will then be the usual prices" [A.22]. The sizes of the reserves and resources are constantly being revised upwards, but with differences depending on who collects the statistics (Table A.2). Resources increase as a result of research and prospecting, that lead to a better understanding of our planet. Resources become reserves when new economically exploitable deposits are discovered, and also as a result of technological progress or higher market prices.

**Table A.2**    Reserves and resources of fossil and nuclear energy carriers, The lifetime of the reserves is based on the conventional reserves of 2006 [BP, BGR] and the mean consumption of the period 2004-2030 (mean IEA prediction). Sources: [a]:[A.1] and [A.8], [b]:[A.4]), [c] [A.23], [d] [A.26], [e] [OECD/IAEA]

| | Reserves 1980 [a] [TWa] | Reserves 1990 [a] [TWa] | Reserves 1998 [a] 1997 [b] 1996 [c] [TWa] | Reserves 2006 [d] [TWa] | Resources 2006 [d] [TWa] | Life in years of conv. reserves |
|---|---|---|---|---|---|---|
| Oil convent. | 118 | 181 | 185[a]/203[b] | 216 | 109 | 36 |
| Oil unconv. | 19 | 68 | 180[b] | 88 | 332 | |
| Gas conv. | 89 | 142 | 170[a]/ 161[b] | 219 | 250 | 48 |
| Gas unconv. | | | 4[b] | | 1852 | |
| Coal | 647 | 1005 | 878[c] | 842[b]/676 | 8101 | 135 |
| Uranium | 39 | 61 | 63[c] | 85[e] | 98 | 65 |
| Total | 912 | 1457 | 1480/1489 | 1450/1284 | 10742 | 284 |

The figures of Table A.2 show that, thanks to coal, there is no threat of a global shortage in the medium term. The critical energy carrier is mineral oil. The so-called "mid depletion point" (that follows the maximum production point and after which no further increase of production is possible) is the date when half the reserves have been used and is estimated by most experts to be 2025-2030. From this date, a large price increase can be expected. The situation is similar for natural gas, with the difference that the reserves are larger and the critical point will not be reached until 2040-2050. There remains the possibility of converting coal into liquid and gaseous fuels and of the technology of separating and storing (sequestration) the resulting $CO_2$. However, the required industrial processes are still at the stage of development and testing. The outlook as regards technical, economic and ecological aspects is therefore still uncertain. Overall, it seems sensible and sustainable, including from the point of view of the reserves, to reduce the world-wide consumption of fossil energy sources more quickly than in the IEA studies (also see section A.7 on this).

The uranium reserves have been calculated for a marginal price of 130 $/kg. This is, however, somewhat elastic, as the cost of uranium has little influence on the price of electrical energy supplied by a nuclear power station. With the occurrence of the modest global increase in nuclear power station capacity expected by the IEA, the critical point is likely to occur relatively late towards the end of the century. There is however, no scope for a large-scale substitution of fossil fuels unless alternative concepts are introduced (4th generation breeder technology, high-temperature reactors and thorium - see [A.5], section 5.6).

### *A.2.1.3 Ethical aspects and environmental protection*

The question of the rate at which these reserves can be allowed to be consumed can be given a good answer from the purely economics point of view (Hotelling's rule and further aspects, e.g. see [A.8]). In addition to this there are ethical considerations and environmental aspects:

- From an ethical viewpoint one must ask whether it is acceptable to extract virtually the whole of this within two or three centuries, the human race's store of energy, that future generations might be dependent on (emergency supplies, raw materials). This aspect should be relativised to the extent that we do not know today whether future generations will in fact be dependent on this energy. But, is this behaviour really responsible?

- The harmful effects of $CO_2$ emissions on the climate, corroborated by scientific studies (e.g. IPCC), justify political action on a global scale in order to produce a drastic reduction in the output of $CO_2$ (section A.7).

As seen today, it is particularly the second aspect that makes it clearly urgent to correct the economic viewpoint by internalising the external costs of present and future damage to the climate.

### A.2.2    Renewable energy

Renewable energy (Fig. A.4) refers to the *natural flows of energy* of a certain magnitude that can be tapped by the use of technical means and with a certain economic cost. Most of these energy sources are not everywhere, but only available or economically exploitable at favourable locations. The various sources will now be considered in respect of their availability.

### *A.2.2.1 Tidal*

Gravitational energy is available to us in the form of *tidal energy*. Tidal friction amounts to about 2.5 TW (i.e. the earth's rotational energy is reduced by 2.5 TWa each year). It is estimated that only about 9% of this is economically usable [A.23]. For economical use, the tidal range must be at least 6 m. Consequently, in relation to future global energy demands (> 25 TW), tidal energy is of very little significance. More on this in, for example, [A.15], [A.16], [A.23].

### *A.2.2.2 Geothermal energy*

The mean natural flow of heat is very small (about 0.06 $W/m^2$). Of this, 30% is from the residual heat of the earth's core and 70% from the decay of radioactive isotopes in the earth's crust. The thermal gradient has a mean value of 1°C per 30 m. Therefore it is above all the geothermal anomalies (volcanoes, geysers) that can be economically exploited. In 2005 the world total installed power was about 28 GW, of which 17 GW was *thermal* (main countries: China, USA, Sweden, Iceland) and 9 GW was *electrical* (main countries: USA, Philippines, Indonesia, Mexico, Italy) with annual production rates (mean power) of 6.6 and 6.5 GW. These are almost all *hydrothermal* uses (hot springs). In the future, the *hot dry rock* process may become more significant [A.14], [A.24].

Geothermal energy is a renewable energy, although over relatively long periods (decades to centuries), so *resources* usable in the short to medium term are indicated. The hydrothermal *resources* with high enthalpy levels, > 150°C, that can be used for generating electricity, amount to about 30 TWa, and the low-enthalpy stocks suitable for heating, about 3200 TWa. How much of this can be called reserves cannot be quantified at present [A.4]. The global installed capacity can presumably be increased to 2 TW (about 100 times the present value), which would cover 5-10% of the future global energy requirement. However, by using the heat pump, this proportion could perhaps be much greater. In some countries the contribution of geothermal energy is very significant.

### A.2.2.3 Solar energy

Solar is the only renewable energy that is already making a significant contribution to meeting global energy requirements, in the form of water power and biomass (2004, about 2 TWa [A.12]).

The supply of solar energy is more than 10000 times as great as the current global energy demand and it is the only energy source, possibly together with nuclear fusion, that can meet mankind's energy requirements in the post-fossil age. The problems around the direct use of solar radiation are related to economics and its low density.

As is shown in Figure A.4, the flow of solar energy is around 173000 TW. Of this energy, 52000 TW are directly reflected back into space as short-wave radiation, while 121000 TW are absorbed, converted and finally, since the earth is in thermal equilibrium, returned into space as long-wave radiation.

About two thirds of the absorbed radiation is stored in air, water and the ground as *low-temperature heat* and, as such, it can be made use of, e.g. by means of heat pumps.

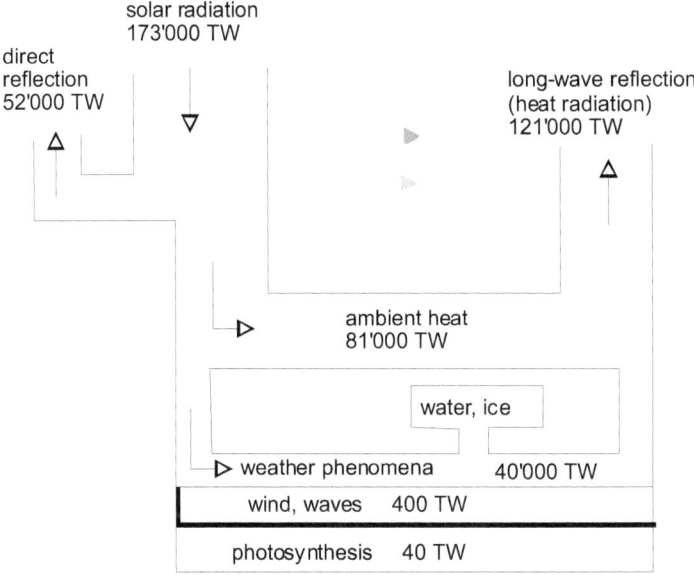

**Figure A.4**   Solar energy balance of the world

The remaining third drives the world's weather by evaporating water and creating differences of pressure and temperature; this energy occurs as *potential energy* (water content of clouds, flowing water and glaciers) and as *kinetic energy* (wind, ocean currents, waves).

Only a fraction, about 40 TW, is absorbed by photosynthesis that creates *biomass* and makes life on earth possible.

In view of its importance, the potential and exploitation of solar energy will now be discussed further.

### A.2.3    The potential and usage of the main types of solar energy

We distinguish between:

indirect (delayed) exploitation of
  - ambient heat (particularly through use of the heat pump)
  - water power (via hydro-electric power stations)
  - wind power, wave power and ocean currents (especially wind turbines)
  - biomass (wood, plants, organic waste) and

direct (immediate) exploitation of
  - solar radiation (solar architecture, collectors, high-temperature heat, photovoltaics).

### *A.2.3.1 Heat pumps*

Heat pumps are a mature technology for making use of low-temperature heat. It enables the temperature to be raised to values suitable for use in space heating and providing hot water. The possible sources of thermal energy are air, ground water, surface water and the interior of the ground. The heat pump makes it possible to use heat from the environment, from waste, and geothermal heat when the source is 100-150 m below ground level.

It is the most mature technology for low-temperature applications for substituting fossil fuels with solar heat and geothermal heat. The potential for ambient heat is enormous, as Figure A.4 shows. Increased use of heat pumps is being impeded in particular by cheap fossil fuels. Also, the fact that about one third of the generated heat must be obtained from high-value energy carriers (electricity, gas) has an inhibiting effect in countries where a high proportion of electricity production is from fossil fuels. For more information about heat pumps, see [A.5], section 5.9

### *A.2.3.2 Water power*

The potential of all flowing water in the world is estimated as 6 TWa (average power), of which about 1-1.5 TWa is economically accessible. [A.23]. In 2005, 0.334 TWa (hydroelectricity) was used. The potential for development is, above all, in the developing countries, but also in the north (Greenland, Canada). The contribution of water power to *meeting the demand for electricity* was, in 2004, 16% globally, 13% in the OECD, 14% in the EU-15 countries, 4% in Germany, 59% in Austria, 56% in Switzerland and 99% in Norway. The use of water power is described comprehensively in [A.5], chapter 4.

### A.2.3.3 Wind and ocean power

The use of wind power has made considerable progress in some countries with favourable wind conditions (Germany, Spain, USA, Denmark) and the associated technology has matured. The economic viability is dependent on the strength and reliability of the wind. Mean wind speeds of at least 5-6 m/s are necessary. These are found mainly in coastal regions and in some mountain regions. If we assume that globally, 2-3‰ of wind and wave energy (400 TWa according to Fig. 1.4) can be economically used, the result is 1 TWa; more can probably be obtained. In 1996 the total installed wind-powered generation in the world was 6 GW [A.23], which increased to 20 GW by 2001 [A.27] and reached 59 GW at the end of 2005 (which corresponds to a growth rate of 30%/yr). In 2006, a level of 75 GW was reached (including 20 GW in Germany), delivering about 12 GW (usage factor 0.16). The offshore installations have a significantly better usage factor. In Denmark it is intended to meet half of the 3.6 GWa electrical energy demand from wind energy by 2030, mainly from offshore installations. In France, 23 GWa of wind energy is expected in 2040. More information about the technology of using wind energy for generating electrical energy is in [A.5], chapter 6.

For possible uses of waves and ocean currents, see [A.29].

### A.2.3.4 Biomass

*Definition.* Biomass consists of materials of organic origin, in other words the material from living beings and organic waste materials (excluding fossil fuels).

*Potential.* The energy in the total biomass of the world is estimated at about 450 TWa [A.24] (1 TWa = 31.5 EJ), with a mean calorific value of about 3500 kcal/kg (referring to fully dry biomass). The annual growth rate is of particular importance, around 60 TWa/a. The conversion efficiency for solar radiation is, on average, about 0.14%, but higher for forests and fresh water (about 0.5%) and highest of all for tropical forests (up to 0.8%). From the chemical point of view, biomass consists of 82% polysaccharides (cellulose and hemicellulose) and 17% lignin [A.15].

*Usage* The technologically exploitable energy potential of biomass in the form of fuels for heating and transport is estimated to be about 6 TW. With a world population of 10 billion people, around 2 TW of this could be obtained from waste. Biomass is thus one of humanity's most significant energy reserves, that could provide about 25% of future demand. Current usage (mainly non-commercial energy) may be (2005) around 1.6 TW. The burning of biomass is $CO_2$-neutral only if reforestation is also undertaken.

*Technical processes for exploitation* (More on this in, for example, [A.15], [A.24]):

- **Physical biological conversion:** This includes *compaction* into bio-fuels (turf, straw, wood waste → briquettes) and the extraction of oil (rape, special oil plants → diesel fuel). If large quantities of plants are cultivated for their oil the danger of monocultures instead of biodiversity must be heeded.

- **Thermo-chemical biological conversion:** Methods to mention include in particular *direct burning* (especially wood) open or in stoves, *gasification/liquefaction* by means of *pyrolysis*, i.e. thermal decomposition of materials of high molecular weight (including used materials such as waste, old tyres, plastics etc.) into smaller molecules and *methanol synthesis* (for obtaining fuel).

- **Biological conversion:** These are low-temperature processes that use microorganisms (fermentation). They include *generating bio-gas* and making *ethanol* from plants with a high sugar content (sugar cane) that is in use on a large scale in Brazil for making fuel.

### A.2.3.5 Solar radiation

*Specific availability of solar energy*
If the 121 000 TW that reaches the earth (Fig. A.4) is distributed uniformly over the earth's surface, the result for a *horizontal surface* is:

$$annual\ mean\ power = \frac{121\,'000\,TW}{510\cdot10^6\ km^2} = 237\ W/m^2 .$$

If we consider only daylight hours (12 instead of 24 h), the result is a doubled mean annual daytime power of 474 W/m². These figures apply to an average latitude, clear weather and at sea level. The effective mean value depends not only on the latitude but also on the climate. In central Europe (often cloudy or misty), half this figure can be expected, 120 W/m².
If we multiply the annual mean power by 8760 h/a, the result is

$$annual\ mean\ energy = 237\ W/m^2 * 8760\ h/a = 2076\ kWh/a\ m^2 .$$

For the reasons stated, this annual energy is not achievable at moderate latitudes because it assumes constant sunshine. The value is exceeded (up to over 2200 kWh/a/m²) near the equator and in desert climates (Sahara, Arizona, Australia etc.).

*Density of solar radiation (global radiation)*
The earth's cross-section is about 127x10⁶ km². Outside the atmosphere this gives

$$extraterrestrial\ radiance = \frac{173\,'000\ TW}{127\cdot10^6\ km^2} = 1360\ W/m^2 .$$

This quantity is also known as the *solar constant*. On the earth's surface (sea level) the radiation density incident on a *plane surface perpendicular to the rays* (without reflections!) is

$$radiance \ at \ sea \ level \ = \ \frac{121'000 \ TW}{127 \cdot 10^6 \ km^2} \ = \ 950 \ W/m^2 \ .$$

In Switzerland the figure of 1000 W/m$^2$ is usually used. This global radiation contains a *direct* component and a *diffuse* component (diffusion from the sky). Under the climate conditions of central Europe, the diffuse component is significant (see [A.5], section 7.4.7).

These figures show, on the one hand, that solar radiation has practically unlimited potential, but also make clear the problems that an economical exploitation of solar radiation encounters. The following lists the most important methods of use:

### Solar architecture

Designing buildings to exploit solar energy can result in a large reduction in the demand for heating energy. This is an option that should be used and promoted much more than at present. For more information, refer to the specialist literature and [A.9], [A.14].

### Flat collectors

Flat collectors can use direct and diffuse radiation. The heat is transferred to a heat carrier (usually water with anti-freeze). High efficiency levels are achieved in low-temperature applications (up to 70% for heating open-air swimming baths, up to 60% for water heating, but only 40-50% for space heating); i.e. the efficiency is strongly dependent on the temperature at which the energy is used. These efficiencies are, however, only achieved for the full irradiation levels and are reduced more than proportionally at lower levels. Collectors are economical at present for heating swimming baths and water heating (especially in summer). For more details, see [A.9], [A.15], [A.14].

### Concentrating collectors

*Parabolic mirrors* (parabolic cylinders or paraboloids) can be used to concentrate direct radiation (diffuse radiation cannot be used this way). This generates high temperatures that can be used for producing process heat and electricity. The mirrors do need to be moved to track the sun. The heat carrier is usually a special oil. Examples of applications are solar cookers for developing countries, solar farms for producing industrial heat and electricity (solar thermal power stations) by using the usual steam process [A.15], [A.21].

Using parabolic cylinders (parabolic troughs) temperatures of 100-400°C are achieved. The water is heated in a tube placed at the focus of the parabola. For higher temperatures, expensive *paraboloids* or *heliostats* are used, which serve both for generating electricity and for carrying out chemical processes [A.24].

### Solar thermal power stations

For generating electricity, *parabolic trough power stations* and *solar tower installations* can be used. By the use of parabolic mirrors or flat mirrors (*heliostats*) that are driven to follow the direction of the sun in one or two axes, the radiation is concentrated on the parabolic trough or the top of a tower. A radiation receiver is located here. The heat is transferred to a working medium (e.g. steam, helium, liquid sodium). Temperatures between 400 and 1200°C are achieved [A.15]. This can drive steam or gas turbines that generate electricity in the conventional way.

Solar thermal power stations are suitable in particular for regions with long-duration sunshine and clear skies, since they are unable to use diffuse radiation. Various pilot plants are in operation around the world [A.9], [A.21]. Two 50 MW power stations are under construction in Spain and about 300 MW is planned. Efficiencies of about 15% have been achieved. For installations of 100 MW, the energy costs are about 0.2 €/kWh [A.21].

*Photovoltaics*

Photovoltaics (PV) makes direct conversion possible from solar radiation into electricity by means of *solar cells*. Commercial-grade *crystalline silicon cells* today give conversion efficiencies of up to 15%. It is hoped that this efficiency can be increased, perhaps approaching 20%. Crystalline cells are still expensive and, in addition, have a poor energy repayment factor because their manufacture requires a large amount of energy. Their technology has made considerable progress and operational experience is good. However, manufacturing costs need to come down and the repayment factor needs to improve.

Commercial PV installations of 500 kW rating using crystalline cells and connected to the grid can at present produce electricity for about 0.6 €/kWh, which is about 10 times the current market price. Consequently they are only marketable where there is direct or indirect subsidy (feed-in compensation). That this exists at all (in 2006 the total installed solar cell capacity in the world reached about 6500 MW [A.18]) is thanks to the unlimited potential of PV and its long-term prospects that are judged favourable in view of expected technological progress, and to considerations of ecology (solar power exchange, eco-electricity). This market is of significance to the extent that it provides an important stimulus for progress in technology. It can be presumed that *large-scale manufacture* will in future lead to considerable cost reductions. For more about PV and photovoltaic power stations, see [A.5], chapter 7.

## A.2.4  Ecological problems

The energy industry, together with the chemical and agricultural industries has the biggest responsibility for putting pollutants into the biosphere. Generating and using energy leads to *pollution of the atmosphere* through emissions from the burning of fossil fuels and biomass, in addition to losses of natural gas. Pollution of the sea from tanker accidents is also not negligible. The *main emissions* are

> oxides of carbon ($CO_2$, $CO$)
> oxides of nitrogen
> compounds containing sulphur
> methane
> ozone.

## A.2.4.1 Predominantly localised effects

The effects of nitrogen oxides, sulphur compounds, CO and ozone in the troposphere on health, plant ecosystems and cultural assets are mainly regional, but to some extent they are also transported by winds over greater distances. Effective measures against the harmful effects have been taken in recent decades in progressive countries by desulphurisation, reduction of nitrogen oxides and catalysers.

Stratospheric ozone destruction (the ozone hole) is rather different, and is mainly caused by emissions of chlorinated fluorocarbons (CFCs) and only peripherally connected with energy use (for more details see, for example, [A.24]).

## A.2.4.2 Strengthening the greenhouse effect

Much more serious consequences for the climate are due to so-called *greenhouse gas emissions*. There is an extensive literature about this. There is a good summary in the fourth IPCC report (Intergovernmental Panel on Climate Change) of 2006-2007 [A.13], and there are other research projects and reports that support its findings. The report comes to the conclusion, more emphatically than earlier reports, that the observable warming of the earth is very probably caused by humans. The most significant greenhouse gas is $CO_2$, that contributes 77% to the strengthening of the greenhouse effect. Other contributions come from methane (15%), CFCs (1%) and $N_2O$ (7%). The concentration of $CO_2$ in the atmosphere is 380 ppm (parts per million), i.e. 36% greater than the pre-industrial value (280 ppm, practically unchanged in the last 10000 years). It is increasing annually by 1.5-2 ppm. Simulations for the year 2100 give, assuming a doubling of the pre-industrial concentration, and depending on the scenario, an increase in the mean temperature of 2 to 6°C (with a probability of 66% for an increase between 2.4 and 4.1°C). Furthermore, the level of the sea could rise by about 50 cm and the number and intensity of extreme climatic events could increase. The uncertainties around this could have positive, but also large negative consequences, e.g. as a result of non-linear feedback effects. A mean temperature increase of 2° corresponds, for medium latitudes, to a shift of the isotherms about 350 km northwards or a height change of about 350 m, but with the possibility of very great regional differences.

The consequences of a *climate tipping point* could be much more dramatic. A deflection, slowing down, or even long-term cessation of the Gulf stream would have catastrophic consequences for western Europe.

The consequences of this climate change, which is very fast in terms of the history of the earth, would certainly be further stress for the already heavily damaged ecosystems (extinction of species, forests). The distribution of water resources and agricultural productivity could be adversely affected world-wide, with increased stress on socio-economic systems, and with particularly great suffering in the regions that are weak and can only adapt slowly or not at all to the changed conditions. Although there can also be winners, the global balance is very negative.

### A.2.4.3 Sustainable development

The term *sustainability* took shape politically mainly following the report of the Brundtland commission in 1987, which defined sustainable development with the words "Sustainable development is development that meets the needs of the present without compromising the ability of future generations to meet their own needs" and called for growth with a view of economics that incorporates social and ecological aspects over time and in different regions.

It therefore concerns the optimising of development in the economy, society and ecology, (the "triple bottom line" of people, planet, profit) and solidarity round the world and into the future. Difficulties come in trying to reach this target, especially from the fact that economic and socio-political thinking is too often short to medium-term, whereas the needs of ecology, particularly in relation to climate problems, require long-term optimisation.

The above paragraph stresses the need for an emphatic global reduction of greenhouse gas emissions, $CO_2$ in particular. The costs of this forceful transformation are quite high, but the costs of a laissez-faire approach would be even higher. It is not a matter of (purely selfish) adaptation to the consequences of climate warming, but above all, of putting a brake on climate warming, reducing it to the extent that is possible and avoiding the enormous costs of social and infrastructural adaptations, that would increase faster than linearly with the increase of mean temperature.

Political initiatives that internalise current and future external costs, and aiming at a way that is compatible with the market economy, are required. In the fight against global environmental damage, it is right in theory to apply capital and knowledge in those areas where the contribution to improving energy efficiency and reduction of $CO_2$ intensity will be maximised (e.g. via the trading of emission certificates). International cooperation in putting this sensible theory into practice is all too often made problematic by the absence of boundary conditions that steer market forces in the right socio-ecological directions, and by political differences. Sometimes this argument is rather an excuse for doing nothing in some regions. Both kinds of effort, global and local (Fig. A.5) are necessary. Seen short-term, regionalisation is suboptimal as regards the global use of capital, but it promotes regional innovation in matters of the environment, and the influence of this could have a global medium and long-term payback (for actions in energy economics see [A.5], section 1.7).

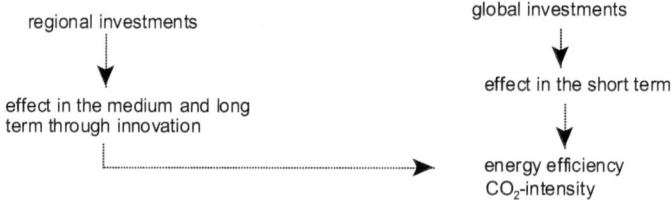

**Figure A.5**   Sustainable investments

## A.3    Energy demand, basics

### A.3.1    General

Before considering the situation regarding the energy sector in Europe (section A.4), of the world (section A.5) and their prospects (sections A.6 and A.7), the energy demand of Switzerland will be analysed as an example of a highly industrialised country. This will enable us to use concrete figures to illustrate the structural aspects of energy demand and to present the factors that determine how it evolves.

### A.3.2    Development and distribution of energy demand

Demand for energy increased greatly as a result of industrialisation. Figure A.6 shows the development of gross energy demand in Switzerland since 1910. *Gross energy* is the annual consumption of local primary energy carriers, plus the import-export difference in primary and secondary energy carriers. This energy demand has increased by a factor of about ten between 1910 and 2000. Since the population has more than doubled over the same period, the per-capita consumption has become about five times as great. Before the second World War, the rate of increase was, on average, scarcely 2%, rose to 7% during the boom period 1950 - 1973 and then swung back to about 2% after the oil crisis of 1973 (note the logarithmic scale). The diagram clearly shows the sharp drops related to the world wars and the economic crisis of the 20s.

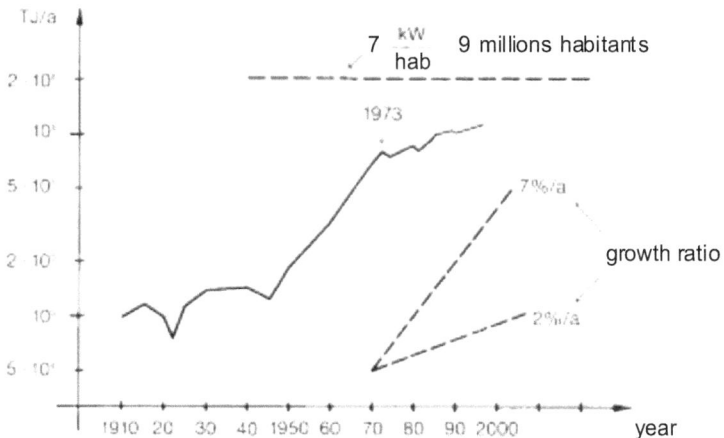

**Figure A.6**   Gross consumption in Switzerland since 1910 (TJ = Terajoule = 1 billion kJ) [A.6]

### A.3.3    Factors that influence energy demand

Figure A.7 shows the development of final energy in Switzerland from 1970 to 2006 and the factors that influence it. These factors are: the resident population, the GDP (gross domestic product) and heating degree days [A.2]. Weather has only a short-term effect. For long-term change, demographic changes and the GDP are the main factors. Table A.3 shows the development of final energy consumption per head in GJ/a/capita and kW/capita.

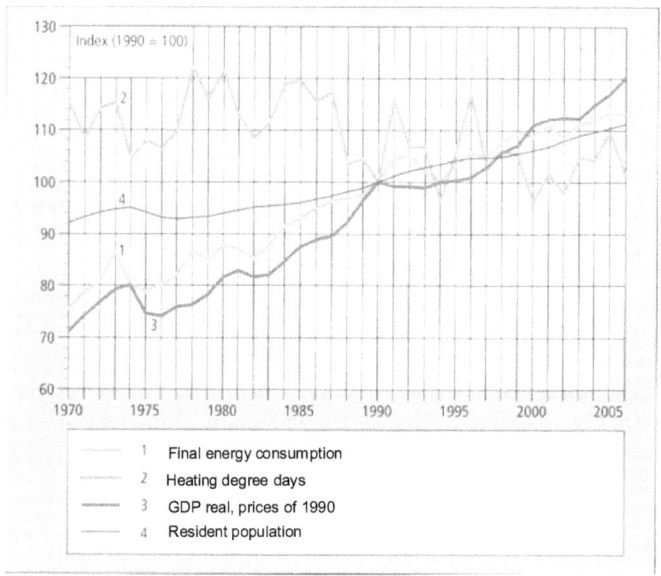

**Figure A.7**  Factors affecting final energy consumption, Switzerland 1970 - 2006, source: [A.2]

**Table A.3**    Final energy consumption per head in Switzerland 1970 to 2006, [A.2]
1 kWh = 3.6 MJ, 1 kWa = 8760 kWh = 31.54 GJ

|  | Final energy consumption TJ/a | Population Million | GJ/a/head | kW/head |
|---|---|---|---|---|
| 1970 | 586 050 | 6.267 | 93.5 | 2.96 |
| 1980 | 697 110 | 6.385 | 109.2 | 3.46 |
| 1990 | 798 510 | 6.796 | 117.5 | 3.72 |
| 2000 | 859 190 | 7.209 | 119.2 | 3.78 |
| 2006 | 888 330 | 7.557 | 117.6 | 3.73 |

### A.3.4    Final energy and losses in the energy sector

For further considerations of sustainability it is useful to distinguish three areas of the use of final energy:

*comfort heating and process heat* that are predominantly obtained from fossil fuels and, to some extent, renewables (geothermal, biomass, solar radiation, ambient heat). Heating by electricity is excluded.

*Fuels for transport* (mainly fossil, some biomass).

*Electricity* (all uses, including heating)

Figure A.8 shows the importance for Switzerland of the three areas as percentages of final energy for the year 2004, (heating 45%, transport fuels 32%, electricity 23%). The three areas are subdivided into the primary energy carriers.

The bar for final energy (100%) also shows the subdivision into the three energy-consuming sectors industry, traffic and others (household, services + agriculture)

Finally, the *losses in the energy sector* are shown, attributable to heat losses in power stations (in Switzerland, effectively the nuclear power stations) and the energy sector's own use of energy. The representation of these losses corresponds to the statistics of the IEA (International Energy Agency) in which not water power but hydroelectricity is recorded. For Switzerland the losses are 28% of final energy. Gross energy consumption in Switzerland is thus 128% of final energy.

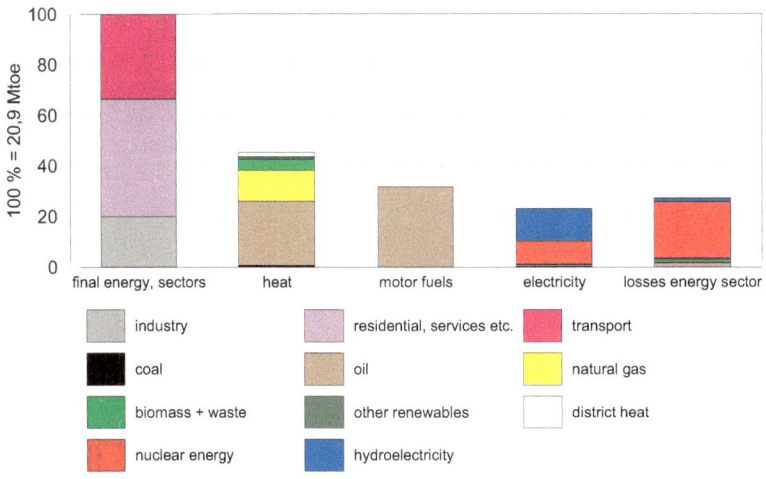

**Figure A.8**  Energy consumption in Switzerland in 2004, as percent of final energy. 100% = 877,290 TJ = 20.9 Mtoe = 243 TWh.  Final energy consists of a) heat (without electricity), b) transport fuels and c) electricity Gross energy is obtained by adding the losses of the energy sector [A.5]

## A.3.5    $CO_2$ emissions and indicators

The emissions of $CO_2$ that result from burning fossil fuels are shown in Figure A.9 for Switzerland in 2004. They total 5.8 t/a/capita and are almost entirely caused by the demand for heating and motor fuels; they are roughly equally distributed, at about 3 t/a/capita, over these two areas. The energy sector (electricity generation and district heating) only makes a minimal contribution to emissions since electricity is 95% produced by water power and nuclear energy (Fig. A.8).

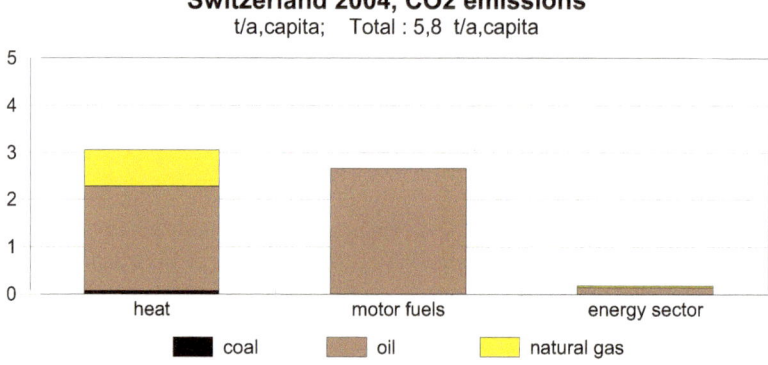

**Figure A.9**    Distribution of $CO_2$ emissions in Switzerland in 2004 [A.5]

For international comparisons it is of interest to relate the emissions to GDP (gross domestic product), taking purchasing power into account. Taking as a reference a GDP of \$10,000 (\$ of 2005), we obtain for Switzerland a figure of 1.75 t $CO_2$/\$10000 (or equivalently 175 g $CO_2$/\$). This figure characterises, better than emissions per head, the sustainability of energy supply of a country in contemplation of climate change. For further analysis, the following relationships are introduced:

$$\alpha \left[\frac{tCO_2}{a, capita}\right] = k \left[\frac{tCO_2}{kWa}\right] \cdot e \left[\frac{kW}{capita}\right]$$

$\alpha = CO_2\text{-}emissions\ per\ year,\ per\ capita$
$k = CO_2\text{-}intensity\ of\ gross\ energy$
$e = gross\ energy\ consumption\ per\ capita$

(A.1)

and

$$e \left[\frac{kW}{capita}\right] = y \left[\frac{10'000\$}{a,\ capita}\right] \cdot \varepsilon \left[\frac{kWa}{10'000\$}\right]$$

(A.2)

$y = specific\ gross\ domestic\ product\ (purchasing\ power\ parity)$
$\varepsilon = gross\ energy\ intensity\ of\ the\ economy$

The specific $CO_2$ emissions can then be expressed as follows

$$\alpha = k \cdot e = k \cdot \varepsilon \cdot y = \eta \cdot y$$

$with\ \eta \left[\frac{tCO_2}{10000\$}\right] = k \cdot \varepsilon = CO_2\text{-}indicator\ of\ energy\ economy$

(A.3)

as the product of affluence indicator $y$, energy intensity $\varepsilon$ and $CO_2$ intensity $k$. To achieve a reduction of the specific $CO_2$ emissions per capita, the wished-for rise in the *affluence indicator y* must be compensated by a large reduction in the *energy intensity $\varepsilon$* and the *$CO_2$ intensity k*. We shall define the product of these two quantities as *$CO_2$ indicator* or *sustainability indicator* $\eta$. The corresponding numbers for Switzerland in 2004 are shown in Table A.4.

**Table A.4**    Indicators for Switzerland, 2004
Dollar 2005, 1 kWa = 8760 kWh = 0,753 toe = 31 540 MJ

| 2004 | $y$ $10^4\$/a/capita$ | $\varepsilon$ $kWa/10^4\$$ | $k$ $tCO_2/kWa$ | $\eta$ $tCO_2/10^4\$$ | $\alpha$ $tCO_2/a/capita$ |
|------|------|------|------|------|------|
| CH | 3.35 | 1.42 | 1.24 | 1.75 | 5.8 |

## A.4    EU-15, final energy and losses of the energy sector

The EU-15 includes the European countries with the strongest economies. As regards population, they have one third of the total OECD population. Their energy consumption and the associated $CO_2$ emissions are shown in Figs. A.10 and A.11, and Table A.5 compares the characteristic indicators with those of Switzerland. The main difference from the corresponding diagrams for Switzerland (Figs. A.8 and A.9) is in the energy sector. Electricity production of many EU countries is based largely on fossil energy sources (e.g. in Germany, Italy, Great Britain, Benelux and Denmark) and the emission of $CO_2$ is correspondingly high (9.4 t/a/capita against 5.8 t/a/capita in Switzerland). However, there are large differences within the EU. For example, France and Sweden have a similar structure to Switzerland, thanks to their electricity generation being based on water power and nuclear power. Outside the EU the same holds for Norway (only water power).

**Table A.5**  Indicators for Switzerland and the EU-15 for 2004. The energy in kWa corresponds to the gross domestic product; $\eta = k \cdot \varepsilon$, $\alpha = \eta \cdot y$

|  | $y$ $10^4$\$/a/capita | $\varepsilon$ kWa/$10^4$\$ | $k$ $tCO_2$/kWa | $\eta$ $tCO_2$/$10^4$\$ | $\alpha$ $tCO_2$/a/capita |
|---|---|---|---|---|---|
| CH   2004 | 3.35 | 1.42 | 1.24 | 1.75 | 5.8 |
| EU-15   2004 | 2.9 | 1.83 | 1.78 | 3.25 | 9.4 |

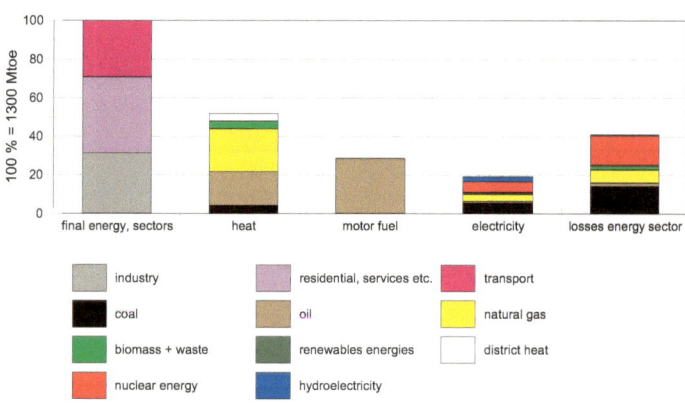

**Figure A.10**   Final energy and losses in the EU-15, 2004, [A.5]

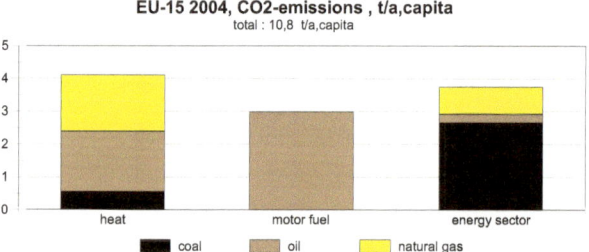

**Figure A.11**   $CO_2$ emissions of the EU-15 in 2004 and their distribution

## A.5   Global demand for final and primary energy

In view of the large gap in GDP per capita that separates the OECD countries from the rest of the world, it makes sense to analyse the OECD and the non-OECD world separately. Apart from the EU-15 that have already been looked at, there are other particularly important countries within these groupings, such as the USA and China and the transition countries, and so they will also be looked at in detail.

### A.5.1   The OECD countries taken together

The OECD countries account for around 18% of the world's population and consume about half of the primary energy. Figure A.12 shows the energy structure of the OECD area for the year 2004 and Figure A.13 the corresponding $CO_2$ emissions.

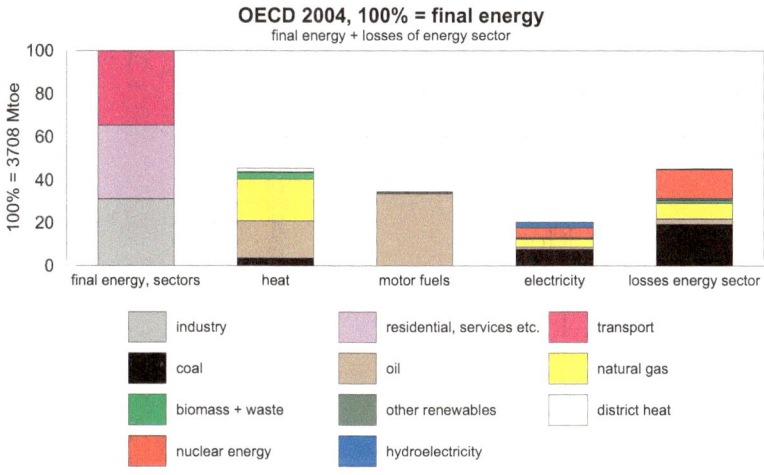

**Figure A.12**   Distribution of final energy and energy sector losses, OECD 2004

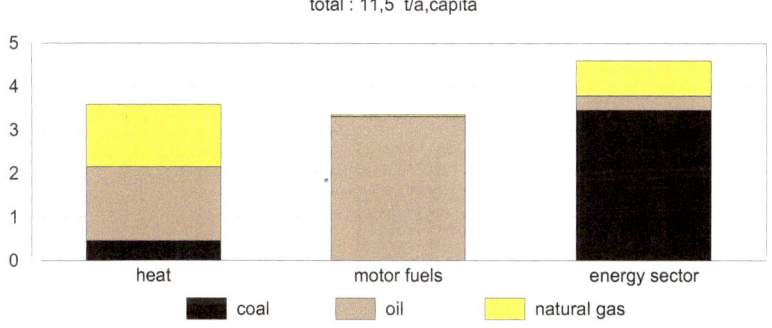

**Figure A.13**   $CO_2$ emissions of the OECD countries taken together, 2004

The structure of the consumption of energy is comparable with that of the EU-15 (Figs. A.10 and A.11). However, although the share taken by electricity is about equally large (20%), it is based on coal and has a lower proportion of renewable energy; consequently it causes greater losses in the energy sector and has primary responsibility for the poor $CO_2$ balance shown in Figure A.13.

*USA*

Since the USA accounts for about 40% of the primary energy of the OECD (and therefore around 20% of world energy demand), their behaviour in conjunction with the demands resulting from climate change is of the highest significance. Figure A.14 shows the structure of the USA's energy consumption and Figure A.15 the corresponding $CO_2$ balance. The per-capita $CO_2$ emissions are 78% higher than the OECD average; the reasons are: the 44% higher GDP (PPP), the 15% higher energy intensity and the 8% higher $CO_2$ intensity of final energy (also see Table A.7). Contributors to these results are the transport sector (taking 40% of the final energy and more significant than heating) and the mainly coal-based production of electricity.

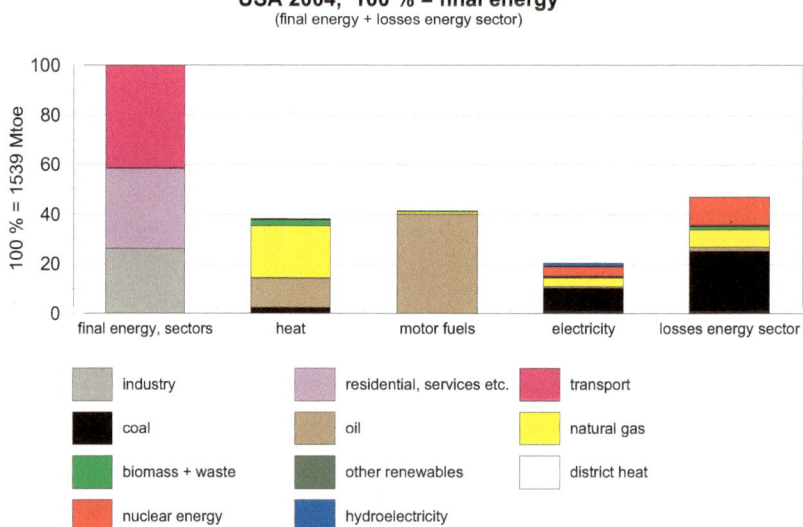

**Figure A.14**   Final energy structure and losses in the energy sector, USA 2004

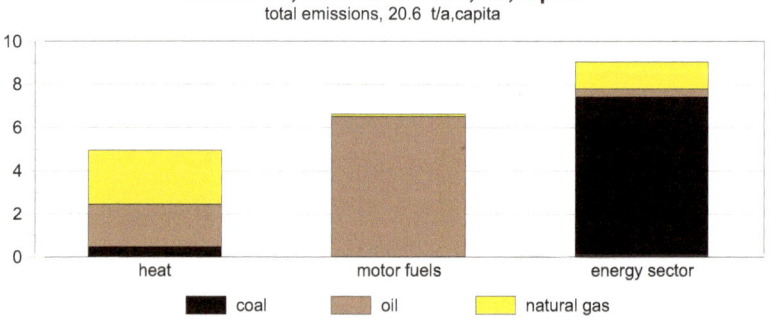

**Figure A.15**   Distribution of $CO_2$ emissions: USA 2004

## A.5.2    Non-OECD countries

The non-OECD countries, with a population (in 2004) of 5.2 billion consume approximately the same amount of energy as the OECD countries (1.2 billion). The energy structure is quite different, as is shown in Figs. A.16 and A.17.

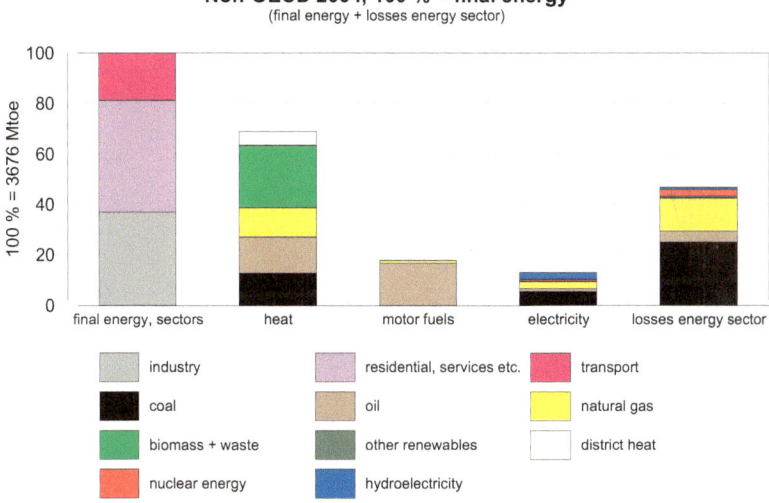

**Figure A.16**  Energy structure of the non-OECD countries taken together, 2004

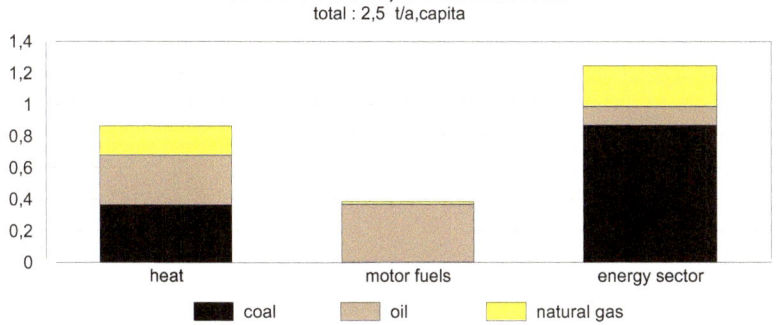

**Figure A.17**  $CO_2$ emissions of the non-OECD countries

Almost 70% of the final energy is for heating purposes. The related $CO_2$ emissions (Fig. A.17) are relatively low, thanks to the large proportion of biomass This will presumably change for the worse in the future.

The same applies to motor fuels since mobility is growing strongly.

The energy sector is a consumer of coal, extremely inefficient, and responsible for half of $CO_2$ emissions, and this despite the fact that electricity is only 13% of final energy.

## *China*

The structure of the energy economy and the associated $CO_2$ emissions are shown in Figs. A.18 and A.19. China's gross energy demand was already 15% of world demand in 2004, or 30% of the demand of the non-OECD countries.

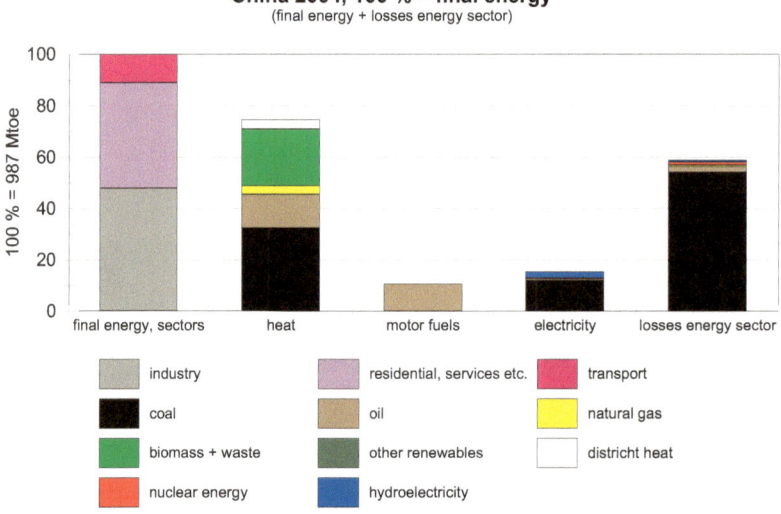

**Figure A.18**   Structure of the Chinese energy economy, 2004

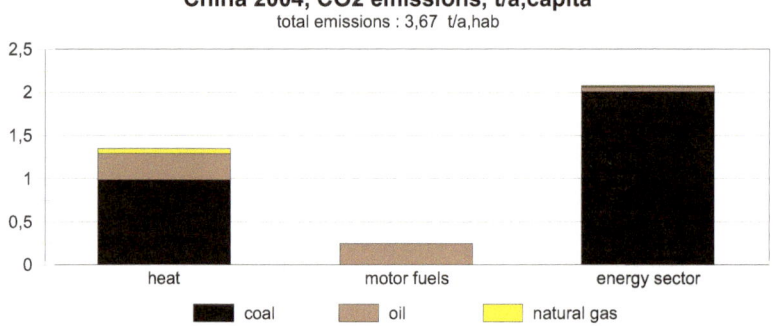

**Figure A.19**   Distribution of $CO_2$ emissions: China 2004

Here too, the demand for heating is 70% of the final energy. The losses of the energy sector are extremely high, reaching almost 60% of final energy demand. The relatively high emissions of $CO_2$ are attributable to the heavy consumption of coal, which accounts for 45% of the heat and nearly 80% of the electrical energy.

### *Transition countries*

The transition countries are all the countries of the former Soviet Union and all European countries that are not members of the OECD. Russia is the most important country of this group (42% of the population, 55% of the GDP and 60% of the energy consumption). It is also influential on account of its large energy reserves, particularly natural gas.

Although the structure of energy use is very similar to that of the non-OECD countries as a whole, or China, a fundamental difference is clear, in that the main energy carrier is natural gas, not coal. (Figs. A.20 and A.21). Nevertheless, the energy intensity of these countries is extremely high (Table A.7), which is explained not by the cold climate but, above all, by the disorganisation and inefficiency of a system that was centrally controlled for years, and by the much too low price of energy.

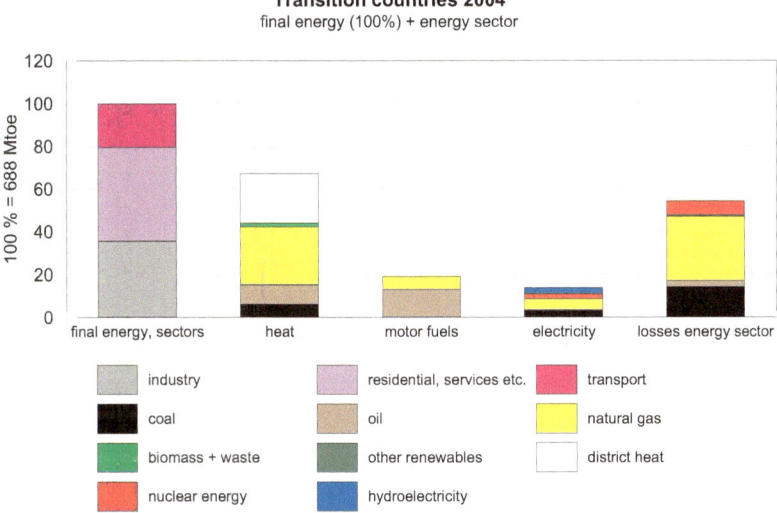

**Figure A.20**   Energy structure of the transition countries, 2004.

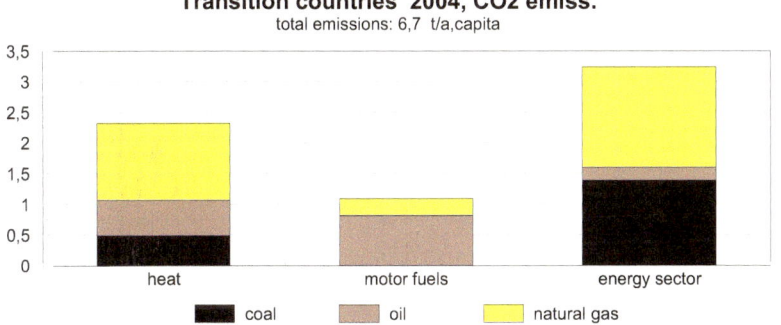

**Figure A.21**   Distribution of $CO_2$ emissions of the transition countries in 2004

## A.5.3    Characteristic indicators

The most important energy-economy indicators, corresponding to the definitions in section A.3.5, are summarised and compared for all the analysed countries or groups of countries and for the world as a whole in Tables A.6 and A.7.

**Table A.6**  Per-capita consumption of final energy (e) and gross energy ($e_g$), and total $CO_2$ emissions for country groups and countries (1 kW = 0.753 toe/a)

| 2004 | e kW/capita | $e_g$ kW/capita | $CO_2$ Mt/a |
|---|---|---|---|
| CH | 3.72 | 4.75 | 45 |
| EU-15 | 4.47 | 6.30 | 4 143 |
| USA | 6.95 | 10.22 | 6 066 |
| OECD | 4.22 | 6.12 | 13 437 |
| World | 1.54 | 2.28 | 26 383 |
| Non-OECD | 0.95 | 1.39 | 12 946 |
| Transition countries | 2.68 | 4.13 | 2266 |
| China | 1.01 | 1.60 | 4764 |

**Table A.7.**  World's characteristic energy-indicators in 2004 ;
y = GDP (PPP) per capita, $\varepsilon$ = intensity of gross energy,  k = $CO_2$ intensity of gross energy;
$\alpha = \eta \cdot y$ = $CO_2$ emissions per capita ; $\eta = k \cdot \varepsilon$ = $CO_2$ indicator (sustainability indicator, climate) : 1  t $CO_2/10^4$\$ = 100 g $CO_2$/\$

| 2004 | y $10^4$\$/a/capita | $\varepsilon$ kWa/$10^4$\$ | k t$CO_2$/kWa | $\eta$ t$CO_2/10^4$\$ | $\alpha$ t$CO_2$/a/capita |
|---|---|---|---|---|---|
| CH | 3.35 | 1.42 | 1.24 | 1.75 | 5.8 |
| EU-15 | 2.9 | 1.83 | 1.78 | 3.25 | 9.4 |
| USA | 4.04 | 2.53 | 2.02 | 5.11 | 20.6 |
| OECD | 2.81 | 2.18 | 1.88 | 4.1 | 11.5 |
| World | 0.91 | 2.48 | 1.83 | 4.54 | 4.2 |
| Non-OECD | 0.49 | 2.83 | 1.81 | 5.12 | 2.5 |
| Transition countries | 0.49 | 5.29 | 1.61 | 8.5 | 6.7 |
| China | 0.66 | 2.42 | 2.3 | 5.57 | 3.7 |

## A.6  Development of the world's population

At the turn of the millennium the population of the world was about 6 billion people. This is a doubling since 1960. Of these, about 1.5-2 bn are materially privi-
leged, while the remaining 4-4.5 bn have a long way to go to catch up. Demographic studies are largely in agreement that the world population, as in Figure
A.22, will increase to at least 8 bn by 2030, and to about 9 bn people by 2050. By
2100, a progressive stabilisation at around 10 bn people is predicted [A.7]. The
population of today's industrialised countries will grow relatively little, mainly
through immigration, and that of the developing countries will increase considerably.

Figure 23 shows the same development, using a different time scale. At the
start of industrialisation in 1850, the whole population of the world was only 1
billion people. The diagram illustrates the unique nature of our times, that will
probably go down in history as the time of the demographic explosion.

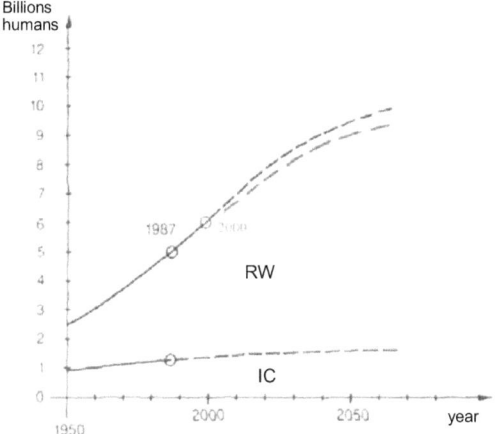

**Figure A.22**   World population since 1950 and forecast
*IC* = current industrialised countries, *RW* = rest of world

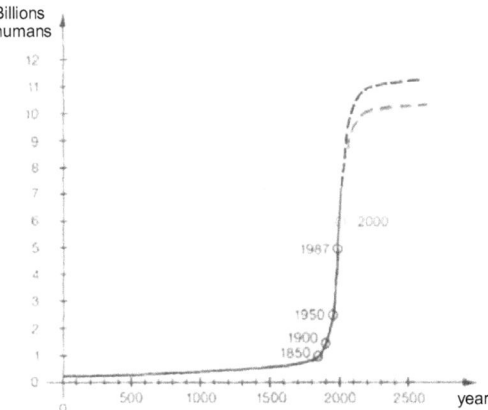

**Figure A.23**   Increase in world population since the start of the present calendar and
forecast further development [A.5]

## A.7    CO₂ emissions and climate protection

### A.7.1    World development of CO₂ emissions, IEA scenarios

For the start of our analysis, *the IEA's alternative scenario* will be used; the *IEA's reference scenario* omits far too many of the boundary conditions required in climate protection.

Figure A.24 shows the energy structure expected by the IEA for the OECD in 2030, and Figure A.25 the corresponding $CO_2$ emissions. The structure and the emissions can be compared with those of 2004 of Figs. A.12 and A.13 (determined from [A.10]).

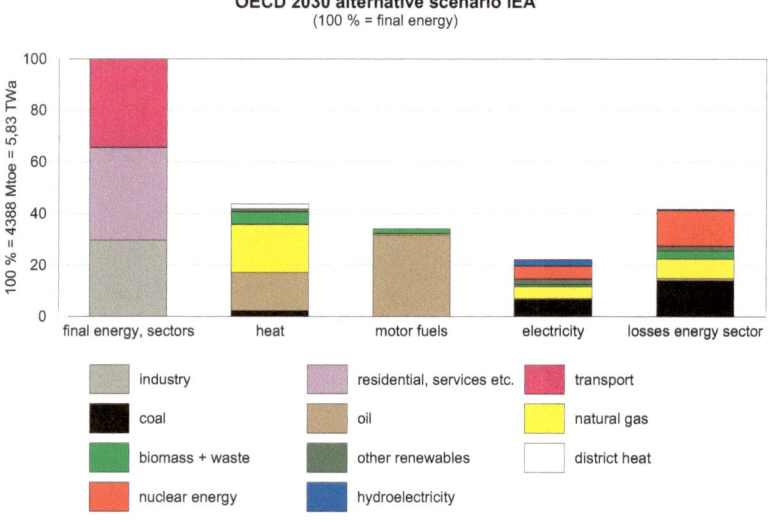

**Figure A.24**  Structure of the OECD energy economy as in the IEA's alternative scenario for 2030. [A.5]

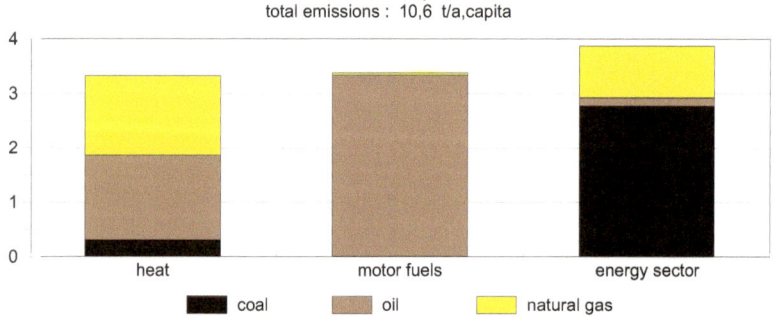

**Figure A.25**  $CO_2$ emissions of the OECD in 2030, as in the IEA's alternative scenario

In parallel with that, Figs. A.26 and A.27 show the energy structure and the $CO_2$ emissions of the rest of the world (non-OECD countries), that can also be compared with those of 2004 (Figs. A.16 and A.17).

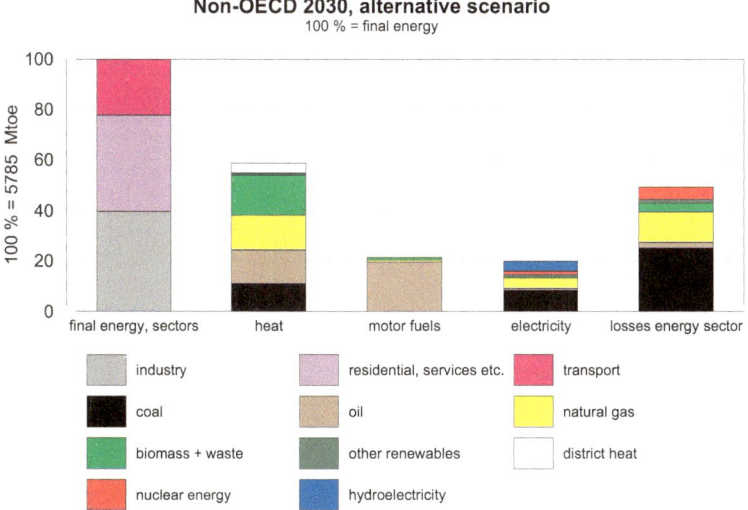

**Figure A.26**   The IEA's alternative scenario for non-OECD countries in 2030 [A.5]

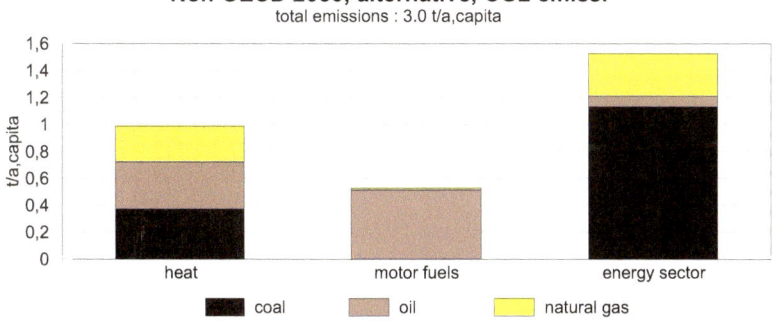

**Figure A.27**   $CO_2$ emissions of the non-OECD countries in 2030, alternative scenario

## A.7.2    World energy indicators for 2030, consequences

Table A.8 summarises the characteristic indicators for 2004 and for 2030 according to the alternative scenario of the IEA (for Switzerland see section A.3 and [A.5]). Notice first that even with the alternative scenario, world emissions of $CO_2$ increase from 27 000 Mt in 2004 to 34 000 Mt in 2030, an increase of 26%. This does not correspond to the targets of climate protection, which demands not an increase but a stabilising of these emissions by 2030. Since the world's population, according to predictions, will increase in this time by about 27%, the specific emissions remain more or less constant at $a = 4.2$ t/a/capita  There is a reduction in the $CO_2$ indicator $\eta$ (and therefore an improvement in sustainability), but this only compensates the increase in GDP/capita.

**Table A.8**  Characteristic indicators from the statistics and the alternative scenario of the IEA for 2030 (calculated from [1.10]) and comparison with Switzerland.
$y$ = GDP (PPP) per inhabitant in $ 2005, $\varepsilon$ = energy intensity (gross energy),
$k$ = $CO_2$ intensity of the gross energy, $\eta$ = $CO_2$ indicator of the economy, $\alpha$ = $CO_2$ emissions per capita ; see also section A.3.5, eqn. (A.3)

| 2004-2030 | $y$ <br> $10^4$\$/a/capita | $\varepsilon$ <br> kWa/$10^4$\$ | $k$ <br> tCO$_2$/kWa | $\eta$ <br> tCO$_2$/$10^4$\$ | $\alpha$ <br> tCO$_2$/a/capita |
|---|---|---|---|---|---|
| Switzerland 2004 | 3.35 | 1.42 | 1.24 | 1.75 | 5.8 |
| Switzerland 2030 | 4.81 | 0.92 | 0.83 | 0.77 | 3.7 |
| OECD 2004. | 2.81 | 2.18 | 1.88 | 4.1 | 11.5 |
| OECD 2030 | 4.47 | 1.43 | 1.66 | 2.37 | 10.6 |
| Non-OECD 2004 | 0.49 | 2.82 | 1.81 | 5.12 | 2.5 |
| Non-OECD 2030 | 1.19 | 1.42 | 1.79 | 2.56 | 3 |
| World 2004 | 0.91 | 2.49 | 1.84 | 4.55 | 4.2 |
| World 2030 | 1.71 | 1.42 | 1.73 | 2.48 | 4.2 |

The $CO_2$ indicator (measure of sustainability) decreases worldwide in this scenario to the ratio 0.55 relative to 2004. It changes from $\eta$ = 4.55 in 2004 to $\eta$ = 2.48 in 2030; this improvement is achieved by:
- the reduction of *energy intensity* (gross energy) with a ratio of 0.57, as a result of technological and organisational progress,
- the reduction of *$CO_2$ intensity* with a ratio 0.94 by means of a (too small) change in the structure of the energy system.

The lack of symmetry between these two factors is striking. To achieve a value of the $CO_2$ indicator that would stabilise world $CO_2$ emissions relative to 2004, the $CO_2$ intensity would also need to be reduced by about the same factor as the energy intensity.

Is it possible to achieve a reduction of this order of magnitude with adequate technological means and political will, and without the forecast effect on the world economy? The answer must be yes, and we shall show in the next section the measures that are essential and entirely feasible for achieving this. The alternative is a climate catastrophe, with its social and political consequences, that would inevitably put a brake on industrial development and with consequences for the economy that could be considerably higher than the radical but necessary measures required of the energy sector to avoid it. A strategy of nothing but adaptation, suggested in some circles, is reckless and irresponsible.

### A.7.3    Climate protection, medium and long-term measures

Various climatological studies [A.13], [A.19] show that to achieve climate protection, the increase in the earth's mean temperature relative to the pre-industrial era must be limited to 2°C. To reach this target it is necessary, as already mentioned, to stabilise $CO_2$ emissions by 2030 (relative to 2004) and to halve them by 2050. (Table A.9, [A.10]).

*The most important action* for most OECD and non-OECD countries is the *reduction of $CO_2$ emission by the energy sector*, which accounts for 45% of all emissions in the world (OECD 40%, non-OECD 50%), primarily in the *generation of electrical energy*. Some European countries are exceptions to this (France, Norway, Iceland, Sweden, Switzerland), as are also most of the countries of central and south America, with almost $CO_2$-free generation of electricity. In 2004 their emission of $CO_2$ was already at or below 250 g $CO_2$/$. If this advantage is retained, the efforts of these countries can be concentrated on fuels for heating and transport.

**Table A.9** Permissible $CO_2$ emissions in 2030 and 2050 for a climate protection scenario, and corresponding emissions per capita and per $ GDP. World starting situation in 2004. The GDP (PPP) is given in $ of 2005: for 2004 and 2030 according to the IEA scenario

|  | Population | $CO_2$ emissionss | Emissions/capita,$a$ | Emissions/$ GDP |
|---|---|---|---|---|
| 2004  World | 6.35 bn | 26400 t | 4.2 t $CO_2$/capita | 460 g $CO_2$/$ |
| 2030  World | 8.1   bn | 27500 t | 3.4 t $CO_2$/capita | 200 g $CO_2$/$ |
| 2050  World | 9     bn | 13200 t | 1.5 t $CO_2$/capita | 60 g $CO_2$/$ |
|  |  |  |  |  |
| 2004  OECD | 1.16 bn | 13400 t | 11.5 t $CO_2$/capita | 410 g $CO_2$/$ |
| 2004  USA | 0.29 bn | 6000 t | 20.6 t $CO_2$/capita | 510 g $CO_2$/$ |
| 2004  EU-15 | 0.38 bn | 3600 t | 9.4 t $CO_2$/capita | 325 g $CO_2$/$ |
| 2004 Non-OECD | 5.19 bn | 13000 t | 2.5 t $CO_2$/capita | 510 g $CO_2$/$ |
| 2004  China | 1.3   bn | 4800 t | 3.7 t $CO_2$/capita | 560 g $CO_2$/$ |

### A.7.3.1    Specific energy consumption and $CO_2$ intensity

The emission of $CO_2$ per person and per year $a$ can be expressed as the product of specific energy consumption $e$ and the $CO_2$ intensity of the gross energy $k$ (Table A.10). At present the gross world consumption is around 2.3 kW/capita, and with a factor of 4 to 5 between the OECD and the non-OECD countries. If the world value is not to exceed 2.5 kW/capita, an emphatic reduction in the specific value of the OECD countries is required (6.14 kW/capita in 2004) in order to compensate the inevitable rise of the value in the non-OECD countries (1.38 kW/capita) which is mainly associated with the expected considerable increase in the GDP of this part of the world. An important step towards this is to increase the energy efficiency (reduce energy intensity) in the industrialised countries, but at the same time to make available to the developing and emerging countries the technologies for increasing energy efficiency.

However, Table A.10 also shows that increasing the efficiency is not enough for reaching the goals of climate protection. It is equally important to achieve a large reduction in $CO_2$ intensity, that applies to OECD and non-OECD countries about equally [A.17].

**Table A.10** Emissions per capita $\alpha$ as the product of gross energy consumption per capita $e$ *(kWa/a)* and the $CO_2$ intensity of the gross energy $k$  (1 kWa = 0.753 toe). World reduction required by 2030 and 2050

|  | Emissions/capita $\alpha$ | Energy consumption/capita $e$ | $CO_2$ intensity $k$ |
|---|---|---|---|
| 2004  World | 4.2 t $CO_2$/a,capita | 2.27 kW/capita | 1.84 t $CO_2$/kWa |
| 2030  World | 3.4 t $CO_2$/a,capita | 2.40 kW/capita | 1.43 t $CO_2$/kWa |
| 2050  World | 1.5 t $CO_2$/a,capita | 2.50 kW/capita | 0.60 t $CO_2$/kWa |
|  |  |  |  |
| 2004  OECD | 11.5 t $CO_2$/a,capita | 6.14 kW/capita | 1.88 t $CO_2$/kWa |
| 2004  USA | 20.6 t $CO_2$/a,capita | 10.20 kW/capita | 2.02 t $CO_2$/kWa |
| 2004  EU-15 | 9.4 t $CO_2$/a,capita | 5.29 kW/capita | 1.78 t $CO_2$/kWa |
| 2004  Switzerland | 5.8 t $CO_2$/a,capita | 4.74 kW/capita | 1.22 t $CO_2$/kWa |
| 2004 Non-OECD | 2.5 t $CO_2$/a,capita | 1.38 kW/capita | 1.84 t $CO_2$/kWa |
| 2004  China | 3.7 t $CO_2$/a,capita | 1.60 kW/capita | 1.84 t $CO_2$/kWa |

### A.7.3.2  Countries with $CO_2$-intense electricity production

The most important countries of this group are the USA, China and, in Europe, Germany, the UK and Italy. Their production of electricity is mainly (China) or heavily dependent on coal and/or oil. Turning away from coal and oil as fast as possible or at least from the current way of using them is the *imperative basic requirement for effective climate protection*. The countries of the OECD should be able to achieve this, given the political will, using their own efforts, and the other countries presumably only with international help. This is made more difficult by the fact that the world demand for electrical energy will approximately double by 2030.

The possible actions and substitutions are:

a)  Large reductions in losses in the energy sector by significantly increasing the efficiency of thermal power stations (combined heat and power, combined cycle generation).

b)  $CO_2$ capture and sequestration in coal and oil-fired power stations; with a significant limitation: the technology is not yet mature, is probably expensive and is not yet fully tested for environmental sustainability.

c)  Use of natural-gas power stations, substitution of coal and oil with natural gas: $CO_2$ emissions, compared to coal, are reduced to about 55% (compared to oil, to about 75%), with a restriction: world-wide gas reserves are limited.

d)  Use of nuclear energy: the power stations do not emit $CO_2$; restriction: reserves of uranium, used in 3rd-generation reactors, are also limited. Use of reactors of the fourth generation is possible, but must be well thought through, technically and politically. Nuclear fusion can only be considered in the second half of the century.

e)  Use of all opportunities to generate electricity from water power; restriction: the potential for this is limited.

f)  Use of wind energy: the technology is mature and, where the wind conditions are favourable, cost-efficient. The potential for this is very great.

g) Use of geothermal and biomass energy. Limitations: geothermal power stations are only suitable in locations with geothermal anomalies. The potential of biomass is limited. Biomass should therefore be used primarily, and provided its use is acceptable, for fuel for transport and heat, with the exception of local combined heat and power.

h) Use of solar thermal energy and photovoltaics. Solar thermal power stations are appropriate only for countries with a low proportion of diffuse light. Photovoltaic generation is currently hindered by its high cost, but as its potential is practically unlimited, its further development must be pursued with determination, including feed-in tariffs as long as this is necessary.

**Remarks:** It is virtually unavoidable that the world demand for electrical energy will double by 2030 for the following reasons, even with an increase of efficiency. It is therefore reckless to rely solely on increased efficiency, even though the increase is absolutely necessary. The use of natural gas and of third-generation nuclear power stations is necessary, although even in the best case this only enables maintenance of their percentage contributions (gas 20% and nuclear energy 15% in 2005), not the replacement of coal and oil-fired power stations (40% and 7% respectively). The same applies for water power (16%). The introduction of the other renewable energy sources f) to h) is therefore essential and must be hugely increased (2% in 2005!). World production of electricity from wind power has already nearly doubled from 2004 to 2006.

### A.7.3.3    The life-cycle cost-benefit, grey energy and electricity exchange

These aspects are often brought into the $CO_2$ balance, making some $CO_2$-free energies or energy uses appear to be less favourable. However, as regards the medium and long-term climate protection goals, this is, for the following reasons, not defensible:

*Life-cycle cost-benefit ratio:* A poor life-cycle cost-benefit factor does have a negative effect on the energy balance (and thus on the economic viability), but not on medium-term climate protection, provided that the energy required for manufacture and transport is also $CO_2$-free, which is required in any case for the medium and long-term target.

*Grey energy:* If the energy used in the manufacture of imported products causes greater emissions of $CO_2$ than that used for producing exported products, the $CO_2$ balance of a country is in theory worsened. However, it is not sensible to include this in the sustainability balance. Each country is ultimately responsible for the energy used in the production of its goods and should put the required climate protection measures in place using its own efforts or in the context of international agreements or with the aid of emissions trading.

*Electricity exchange:* For countries such as Switzerland that export or import large amounts of electrical energy at various times, the $CO_2$ emissions of the power stations required for the production of the exchanged energy can lead to a significant distortion of the $CO_2$ balance. In Switzerland, because of the importing of coal-intensive power the $CO_2$ burden of electricity consumption, while still small, is calculated as about seven times higher than the low burden from domestic electricity production. However, as with grey energy, it is not useful to include this in the $CO_2$ balance of the countries concerned. Exports and imports of $CO_2$-

heavy electricity must be determined and appropriately penalised by market means (emissions trading) or by $CO_2$ taxes. The resulting impediment to the purely economics-based trade in electrical energy is fully in line with the idea of climate protection.

### A.7.3.4  Countries with nearly $CO_2$-free electricity production

These countries have the great advantage of already having solved the most important emissions problem: Switzerland and Sweden with a mix of water power and nuclear energy, Norway with water power, Iceland with water power and geothermal power, France with mainly nuclear energy and Latin America mainly with water power. The first priority is naturally given to maintaining the freedom from $CO_2$ in the electrical energy sector, which can only be achieved by the measures d) to h). These countries already have the basic prerequisite and so they can be pioneers in the development and application of new technologies for limiting emissions in the areas of transport and heating. The areas will now be analysed.

### A.7.3.5  Transport

Emissions are almost entirely caused by the use of fuels obtained from mineral oil, which accounted for around 22% of world $CO_2$ emissions in 2004, and increasing (OECD 29%, non-OECD 15%). In the short term the increase can be kept within bounds by improvements in efficiency (and the associated reduction of $CO_2$ emissions per km travelled), by hybrid technology (using a secondary electric motor) and by partial substitution of petrol and diesel by biofuels (where ecological considerations are important and require further examination).

However, in the medium and long term, protection of the climate is possible only with a paradigm shift. The future must belong to the hybrid solution, with electric motors as the primary drive units and combustion engines as secondary modules for improving the range (as far as possible using biofuels provided that their production involves a good $CO_2$ balance and is compatible with the world requirement for food production). At least 75% of vehicles are driven for less than 50 km per day. The battery supplying the electric motor can then be charged overnight, which will at least enable urban traffic to be largely $CO_2$-free. The production of electrical energy in sufficient quantity and as far as possible free of $CO_2$ together with powerful batteries are the prerequisites for this change which will therefore not happen immediately, at least not globally. Using a fuel cell to supply the electric motor could also make a significant contribution in the long term; for this, however, hydrogen made using $CO_2$-free energy is required.

The change is also imperative from purely economic considerations. The fuel for a very efficient mid-range car with a consumption of, for example 6 l petrol per 100 km (46 mpg, emissions about 140 g $CO_2$/km) currently costs about 8.4 euro/100 km at a price of 1.4 euro/litre. The energy content of 6 l petrol is 52.6 kWh and with an average efficiency of 20% this gives a mechanical power (useful energy) of about 10.5 kWh/100 km. The price of the mechanical drive energy is therefore today already at least 0.8 €/kWh and rising. With the electric motor and including the battery and the power electronics, an average efficiency of at least 65% can be achieved, giving, for the same useful mechanical energy of 10.5 kWh, an electricity consumption of at most 16 kWh/100 km. To achieve the same energy costs as with the combustion engine, the electrical energy from the mains socket for recharging the battery can cost 0.5 €/kWh and more. The corresponding comparison with diesel fuel gives 0.45 €/kWh. Consequently, the mains electricity "fuel" is already much cheaper in most countries than petrol or diesel and the time is not far off, when photovoltaic electricity will be cheaper, even from small installations. Even when the extra power required for the heavier vehicle (because of the battery) is allowed for, there is a clear advantage in the energy costs.

### A.7.3.6 Heating

Energy for heating (excluding electricity) caused 33% of world $CO_2$ emissions in 2004. This is about the same figure for OECD countries (31%) and the rest of the world (35%).

*Heating for comfort* with the lowest emissions possible should not cause particular difficulties, given the appropriate promotion. Suitable means are: solar architecture and good insulation (the Minergie standard), solar collectors, biomass (wood) and, not least, district heating (combined heat and power) and, above all, the heat pump (for using environmental and geothermal heat). For the latter, the restriction applies that the most $CO_2$-free electricity possible should be used (in Switzerland and France conditions are almost ideal for this), giving a new dimension to the importance of an adequate and low-$CO_2$ production of electricity. With a modern heat pump 25-30% of the heating energy is supplied by the electricity.

For *process heat* the contribution from fuels should be reduced in favour of electricity (the most $CO_2$-free possible) and the most emissions-free possible in industry by more efficient processes and the use of biomass, above all in the form of waste.

Let it be emphasised again that in the use of biomass, this is not $CO_2$-neutral unless woodland clearance is compensated by new growth (preservation of forests and the rain forests in particular).

# Bibliography of the annex

[A.1]   BP-Amoco-Statistical Rewiew 1998, www.bpamoco.com (1999)
[A.2]   Bundesamt für Energie: Schweizerische Gesamtenergiestatistik 2006, Sonderdruck, Bern (2007)
[A.3]   Bundesamt für Energie: Schweizerische Gesamtenergiestatistik 2000, Bulletin SEV/VSE, Bern (2001), 16
[A.4]   Bundesanstalt für Geowissenschaften und Rohstoffe, Hannover: Reserven, Ressourcen und Verfügbarkeit von Energierohstoffen 1998, Stuttgart (1999)
[A.5]   Crastan V.: Elektrische Energieversorgung, Band 2, 2.Auflage, Springer Verlag, Berlin, Heidelberg, (2009)
[A.6]   Crastan V.: Die Energiepolitik im Spannungsfeld von Ökologie und Fortschritt. Bieler Publikationen, Gassmann, Biel (1989).
[A.7]   DSW: Weltbevölkerung, Entwicklung und Projektionen, Deutsche Stiftung für Weltbevölkerung, www.weltbevoelkerung.de (2002)
[A.8]   Erdmann G.: Energieökonomik.. vdf Hochschulverlag, Zürich / Teubner Stuttgart (1992)
[A.9]   Goetzberger A., Voss B.: Sonnenenergie. Teubner (1997)
[A.10]  IEA, International Energy Agency, World Energy Outlook (2006)
[A.11]  IEA, International Energy Agency, Key World Energy Statistics from the IEA, Paris, www.iea.org/stats (2006)
[A.12]  IEA, International Energy Agency, Key World Energy Statistics from the IEA, Paris, www.iea.org/stats (2005)
[A.13]  IPCC (Intergovernmental Panels on Climate Change), 4.Bericht (2007)
[A.14]  Kaltschmitt M., Wiese A.: Erneuerbare Energien, Springer, Berlin, Heidelberg (1995)
[A.15]  Kleemann M., Meliss M.: Regenerative Energiequellen, Springer, Berlin, Heidelberg (1988)
[A.16]  Kugeler K., Phlippen P. W.: Energietechnik, Springer, Berlin, Heidelberg (1993)
[A.17]  Crastan,V.: Klimawandel, eine Analyse der weltweiten Energiewirtschaft, Bulletin SEV/VSE, Bern (2007), 19; Notwendige Maßnahmen (2008), 8
[A.18]  Häberlin H. Photovoltaik, VDE/AZ- Verlag, Aarau (2007)
[A.19]  Stocker T.: Die Erde im Treibhaus, Bulletin SEV/VSE, Bern (2007), 1
[A.20]  Schweizerischer Energierat: Schweizerische Gesamtenergiestatistik 1997, Sonderdruck Bulletin SEV/VSE (1998), 16
[A.21]  Solarthermische Kraftwerke. www.fv-sonnenenergie.de (2002)
[A.22]  Spreng D.: Graue Energie. vdf Hochschulverlag, Zürich / Teubner, Stuttgart (1995)
[A.23]  WEC, World Energy Council, Statistik 1996, www.worldenergy.org/wec (2000)
[A.24]  Wokaun A.: Erneuerbare Energie. Teubner, Stuttgart, Leipzig, (1999)
[A.25]  www.bmwi.de/BMWi/Navigation/energie.html (2008)
[A.26]  www.bgr.bund.de (2006)
[A.27]  www.windpower.dk (2002)
[A.28]  www.ec.europa.eu/eurostat (2007)
[A.29]  www.energie-fakten/PDF/meeresenergie.pdf (2005)

# Index

## A

Africa 9, 18, *30*
Albania 38
Algeria 30
Angola 30
Argentina 26
Armenia 38
Australia 30
Austria 10
Azerbeijan 38

## B

Bahrain 34
Bangladesh 22
Belarus 38
Belgium 10
Benin 30
Biomass 6, 9, 20, 25, 32, 51, 55, 56, 57, 65, 71, 81, 83
Bolivia 26
Bosnia-Herzegovina 38
Botswana 30
Brazil 3, 26, 44, 58
Brunei 22
Bulgaria 10, 38

## C

Calorific value 57
Cambodia 22
Cameroon 30
Canada 14, 56
Central and South America 2, 18, *26*
Chile 26
China 2, 3, 9, *18*, 43, 44, 54, 72, 74, 79, 80
Climate change 1, 7, 17, 66, 70
Climate problems 1, 6*2*
- $CO_2$ emissions 1, 2, 3, 7
- Copenhagen 1
- Kyoto 1
- Climate hysteria 1
Climate protection 1f, 5f, 21, 42, 44, 76f
Columbia 26
Combined-cycle generation 8, 80
Combined heat and power 8, 9, 48, 80, 81, 83

Congo 30
Costs 5, 48, 53, 54, 60, 62, 78, 83
Costa Rica 26
$CO_2$ emissions 1, 2, 3, 4, 6, 7, 24, 42, 43, 44, 54, 62, 66f, 79
$CO_2$ intensity
- Africa 20, 32
- Central and South America 20, 28
- China 20
- EU-15 16
- EU-27 12
- India 20
- Japan 16, 20
- OECD 15, 20
- Middle East 20, 36
- Rest of Asia/Oceania 20, 24
- Transition countries 40
- USA 16, 20
- World 20
$CO_2$ sustainability 2, 42
- Africa 21, 33
- Central and South America 21, 29
- China 21, 43
- EU-15 17
- EU-27 13
- India 21
- Japan 17, 21
- OECD 16, 21, 42
- Middle East 21, 37
- Rest of Asia/Oceania 21, 25
- Transition countries 41
- USA 17, 21, 42
- World 2, 3, 21
Croatia 38
Cuba 26
Cyprus 10, 38
Czech Republic 10, 14

## D

Dem. Rep. Congo 30
Denmark 10, 57, 68
Developing countries 1, 5, 56, 59, 75
Dominican Republic 26